About Island Press

Island Press is the only nonprofit organization in the United States whose principal purpose is the publication of books on environmental issues and natural resource management. We provide solutions-oriented information to professionals, public officials, business and community leaders, and concerned citizens who are shaping responses to environmental problems.

In 2006, Island Press celebrates its twenty-first anniversary as the leading provider of timely and practical books that take a multidisciplinary approach to critical environmental concerns. Our growing list of titles reflects our commitment to bringing the best of an expanding body of literature to the environmental community throughout North America and the world.

Support for Island Press is provided by the Agua Fund, The Geraldine R. Dodge Foundation, Doris Duke Charitable Foundation, The William and Flora Hewlett Foundation, Kendeda Sustainability Fund of the Tides Foundation, Forrest C. Lattner Foundation, The Henry Luce Foundation, The John D. and Catherine T. MacArthur Foundation, The Marisla Foundation, The Andrew W. Mellon Foundation, Gordon and Betty Moore Foundation, The Curtis and Edith Munson Foundation, Oak Foundation, The Overbrook Foundation, The David and Lucile Packard Foundation, The Winslow Foundation, and other generous donors.

The opinions expressed in this book are those of the author(s) and do not necessarily reflect the views of these foundations.

Corridor Ecology

Corridor Ecology

The Science and Practice of Linking Landscapes for Biodiversity Conservation

Jodi A. Hilty, William Z. Lidicker Jr., and Adina M. Merenlender

Foreword by Andrew P. Dobson

Washington • Covelo • London

Library of Congress Cataloging-in-Publication data.

Hilty, Jodi A.
Corridor ecology : the science and practice of linking landscapes for biodiversity conservation / Jodi A. Hilty, William Z. Lidicker Jr., and Adina M. Merenlender ; foreword by Andrew P. Dobson.
    p.   cm.
Includes bibliographical references and index.
ISBN 1-55963-047-7 (cloth) — ISBN 1-55963-096-5 (pbk.)
1. Corridors (Ecology) I. Lidicker, William Zander, 1932– II. Merenlender, Adina Maya, 1963– III. Title.
QH541.15.C67H55 2006
577.27—dc22   2006005646

British Cataloguing-in-Publication data available.

Printed on recycled, acid-free paper ✪

Interior design by Karl Janssen
Cover photo by: William B. Karesh, D.V.M.

Manufactured in the United States of America

10 9 8 7 6 5 4 3 2

For generations to come—

*may wildlife and wild places be a cherished
and significant part of their futures*

Contents

Foreword

The nocturnal photographs of terrestrial light emissions observed from space provide a sharp image of how we have fragmented the earth's natural habitats. Huge areas of light demarcate the presence of cities, and they are linked by chains of light that illuminate the paths of the superhighways and roads that break up the remaining patches of natural habitat. If we look at Europe, the eastern United States, Southeast Asia, and West Africa, the areas of darkness are dwarfed by the endless fission and fusion of towns, cities, and villages connected by a cobweb of roads and houses. If we look at the Amazon; the Poles; central Africa; and the northern parts of Russia, China, Canada, and Australia, we can still see areas of continuous open habitat, only occasionally interrupted by dots representing the roads that have made the first vital cuts through these still largely intact wilderness areas. Even the edges of the oceans have small necklaces of lights around coasts, islands, and oil exploration platforms. These nocturnal images illustrate how efficiently humans have fragmented the natural world by subdividing it and competing with each other to own and exploit increasingly smaller and more valuable patches of land.

The stunning sequences of views of Earth from space are perhaps the one aspect of NASA's space program that has had a broad and useful impact on the human psyche. The more distant diurnal images first glimpsed by astronauts in the 1960s still haunt us; our planet is the only celestial object that seems to be divided into areas of land and areas of water. The presence of the latter allows the huge diversity of life that makes Earth the most biologically diverse place in the universe. We have found no signs of life anywhere else in the universe, despite the irony that we spend more money looking for life on other planets than on exploring and conserving the life we have on Earth. NASA also provides images of land cover from space. The occasional (and too temporally erratic) images of forests it has provided over the last quarter century provide significant evidence that all of the world's natural forests are rapidly declining in area. Simultaneously, they are being transformed into smaller and more isolated patches by continuous fragmentation, primarily as a consequence of road building and conversion of land to agriculture. NASA provides very occasional glimpses that suggest similar processes are impacting mangrove swamps, estuaries, savannas, prairies, and even the arctic tundra (Dobson 2005).

If we travel to India and on to China and Southeast Asia, we encounter some of the world's oldest modified landscapes. The civilizations that modified these landscapes have produced a vast agricultural landscape that contains only small isolated patches of natural habitat. Even the remaining patches are rapidly eroding species: The tigers, elephants, and monkeys that played such a key role in the legends that underlie the Hindu and Buddhist religions now persist in only a decreasing number of isolated areas. The landscape of central India illustrates the deep tragedy and several ironies of habitat loss. The central tragedy is that habitat conversion has occurred in order to create the agricultural land needed to feed India's human population. In the initial stages of conversion, agriculture was relatively inefficient, and small areas of land were required to feed each family. Frequently, areas that had been converted to agriculture were later abandoned, particularly in places with poorer-quality soils or significant variation in water supply. This allowed only a nomadic human population to exist in a landscape that persisted as a mosaic of patches of pristine, recently converted, and abandoned and recovering lands. As agriculture developed and humans learned new techniques that increased the viability of the soil, and as they developed means of storing and moving water from rivers and lakes with more persistent supplies, large areas of land could be permanently converted to agriculture. Ultimately, the landscape was converted from a mosaic of patches in different states of conversion and recovery to one in which huge areas are dominated by agriculture interspersed with occasional villages, towns, and cities. Relatively simple models of this process of habitat conversion suggest that the proportion of land that remains as pristine, undisturbed habitat will decline exponentially with the length of time that it can persist under agriculture (Dobson, Bradshaw et al. 1997).

Here we hit a curious irony: The intensification of agriculture under the green revolution produced a huge increase in the productivity of land converted to agriculture (Conway 1997). That in turn has led to a significant reduction in the demand for new land for agriculture. The Millennium Ecosystem Assessment recently estimated the extent of land used by agriculture in 1960 and in 2000 (Millennium Ecosystem Assessment 2003). Although the area of land used by agriculture has increased, the amount of land used to produce the world's current food crop is less than half of what it would be if food were still produced with the efficiency that characterized 1960s agriculture. If we had not had the green revolution, the land needed to feed the current population of India and Southeast Asia would exceed the total land in that area. Hopefully, increases in agricultural efficiency will continue, though they need to be developed in ways that are less dependent on fossil fuels and less wasteful of fertilizers. (Fertilizers illustrate the hidden costs of agriculture, particularly when their runoff concentrates in nearshore marine areas, where they significantly impact fisheries and lead to huge eutrophication problems [Turner and Rabalais 1994, Vitousek, Aber et al. 1997a, b].)

Unfortunately, we have only recently begun to appreciate that the value of the land we have converted may be strongly dependent on services provided by the other species that used to live there, or by the services of species that survive in adjacent patches of unconverted habitat (Kremen, Williams et al. 2002, Dobson 2005). Maximizing the diversity of species we can save and minimizing the loss of ecological services are major conservation challenges of the twenty-first century. The twin processes of habitat conversion and fragmentation have caused the local extinction of many species. Each loss reduces the efficiency of the ecosystem functions undertaken by those species, which reduces the ecosystem services provided to the landowner and the local economy, incrementally reducing the value of the converted land.

Increasingly, we need to focus our attention on how we can move from land conversion to habitat restoration (Dobson, Bradshaw et al. 1997, Simberloff, Doak et al. 1998, Dobson, Ralls et al. 1999). Central to this is understanding how different species and ecological communities persist in a fragmented environment. Ultimately, we have to recognize that this leads us to the toughest theoretical and empirical scientific question of the twenty-first century: How are natural communities organized, and how can we reassemble them in ways that allow them to undertake the full diversity of ecosystem functions that provide the services on which all life on Earth depends? The scientific complexity of this problem easily equals the complexity of the "deep" scientific problems of physics such as the structure of atoms and the birth of the universe. However, there is a significant difference: In many ways, the solutions to the perceived deep problems of physics and chemistry are like those of the *Glass Bead Game* played by Hermann Hesse's fictitious elite protagonists; they will be comprehensible only to a handful of scientists and will have limited utility to the quality of human life. Ultimately, the urgency of their solution is determined by the career trajectories of those who understand them. Ironically, the problems of physics may also involve fewer interacting variables and populations than those of ecology. Attempting to understand the assembly and destruction of ecological communities and food webs that live in small woodlots or lakes may ultimately prove as mathematically taxing as understanding the birth and death of galaxies. On the other hand, developing a deeper understanding of ecological communities has an urgent time line. If we fail to develop such an understanding within the next twenty to fifty years, there will be very few intact communities to use as models for the restored communities we hope to build. This will significantly reduce our ability to maintain the processes that make a major contribution to the viability of all human health and economic welfare.

This book provides a comprehensive introduction to the science and the technical and political logistics involved in this process. The chapters of this book provide key insights into how fragmentation affects the viability of diverse varieties of species. The key solution to this crisis is to develop corridors that reconnect the surviving patches of natural habitat in the landscape. Jodi Hilty, William Lidicker, and Adina Merenlender have collated and organized a major overview of the science and practice of connecting landscapes. The lessons presented here are central to developing an understanding of how we reverse the processes of habitat loss and fragmentation. These lessons need to be applied both in areas that have already been converted and in the dark areas of the nocturnal earth where there may still be time to modify the ways in which we convert the natural landscape.

Andy Dobson
Princeton, New Jersey, and Rajasthan, India
November 2005

References

Conway, G. 1997. *The doubly green revolution: Food for all in the 21st century.* London: Penguin.

Dobson, A. P. 2005. Monitoring global rates of biodiversity change: Challenges that arise in meeting the 2010 goals. *Philosophical Transactions of the Royal Society B: Biological Studies* 360:229–244.

Dobson, A. P., A. D. Bradshaw, and A. J. M. Baker. 1997. Hopes for the future: Restoration ecology and conservation biology. *Science* 277:515–521.

Dobson, A. P., K. Ralls, M. Foster, M. E. Soulé, D. Simberloff, D. Doak, J. A. Estes, L. S. Mills, D. Mattson, R. Dirzo, et. al. 1999. Corridors: Reconstructing fragmented landscapes. In *Continental conservation: Scientific foundations of regional reserve networks,* eds. M. E. Soulé and J. Terborgh, pages 129–170. Washington, DC: Island Press.

Hesse H. 1943. *The Glass Bead Game,* (Magister Ludi). Translated from the German by Richard & Clara Winston, Penguin Books, Harmondsworth, UK.

Kremen, C., N. M. Williams et al. 2002. Crop pollination from native bees at risk from agricultural intensification. *Proceedings of the National Academy of Sciences* 99 (26). 16812–16816.

Millennium Ecosystem Assessment. 2003. *Ecosystems and human well-being: A framework for assessment.* Washington, DC: Island Press.

Simberloff, D., D. Doak et al. 1998. Regional and continental restoration. In *Continental conservation: Scientific foundations of regional reserve networks,* eds. M. Soulé and J. Terborgh. Washington, DC: Island Press.

Turner, R. E., and N. N. Rabalais, 1994. Coastal eutrophication near the Mississippi river delta. *Nature* 368:619–621.

Vitousek, P. M., J. D. Aber et al. 1997a. Human alteration of the global nitrogen cycle: Causes and consequences. *Issues in Ecology* 1:1–15.

Vitousek, P. M., J. D. Aber et al. 1997b. Human alteration of the global nitrogen cycle: Sources and consequences. *Ecological Applications* 7 (3): 737–750.

Preface

The main tasks of conservation biology are to provide the intellectual and technical tools to enable society to anticipate, prevent, and reduce ecological damage, and to generate the scientific information on which effective conservation policies can be designated and implemented.

Michael Soulé and Gordon Orians, 2001

The three of us began our collaboration in 1996 at the University of California, Berkeley. Jodi was designing her graduate research on wildlife corridors; Adina was her thesis advisor, and Bill served on her thesis committee. The research needs on the subject of connectivity were somewhat daunting, and the approaches to quantifying the actual use of corridors at a landscape scale were in their infancy. It was clear to us that for corridors to be a long-term conservation tool, they would need to function for more than one species and ideally for whole communities. Therefore, the research focused on using passive methods to monitor the use of corridors by multiple species, rather than tracking individual animals, which was more commonly done at the time. Jodi's dissertation ultimately demonstrated that use of corridors by native carnivores was influenced by corridor width (Hilty and Merenlender 2004).

Over the course of the field research, which was publicized in local newspapers, we received requests for information about wildlife corridors from local nonprofit conservation organizations, individuals, and resource agencies. We drafted a report for Circuit Riders, Inc., a local nonprofit environmental organization, that briefly reviewed the state of research on wildlife corridors, which we distributed. We soon realized that the notion of wildlife corridors had been adopted wholeheartedly and many people were eager for guidelines to help design and manage them successfully. Clearly, an up-to-date resource was needed that could distill the science for practical application. We decided to write this book to help fill that evident need.

We believe that part of the strength of this book comes from the different perspectives that each of us brings to the task, both because of our different expertise and because we are at different stages in our professions. Bill has more than fifty years of academic experience in basic science departments, which helped ensure a sense of historical depth, as well as a chronology of theory and research. Adina is in the middle of her career as a conservation biologist and works to extend applied science to multiple user groups, which requires translating scientific information to lay audiences. Jodi is early in her career as an applied conservation scientist and is well informed on existing conservation projects and recent advancements reported in the literature; her energy, enthusiasm for the subject, and leadership kept us on track and made this book possible.

We committed to writing this book with one voice. In order to divide the workload practically, we each took the lead on an equal number of chapters initially, but through the process of consolidating sections and multiple reviews, we all had substantial input on all chapters. This book would not have been possible without all of us, and therefore the authors are listed in alphabetical order.

We owe thanks to the many people who helped us develop as scientists and conservationists, ultimately leading us to this book. These include our many mentors, from parents to teachers and from advisors to students. In particular, we want to thank W. F. (Bill) Laurance, who, as a graduate student, was most influential in getting Bill actively involved in academic conservation biology at a time when conservation research was not considered appropriate for a basic science department. Glenda and Bill Hilty, Bill Weber and Amy Vedder, Dale Lewis, and Martha Schwartz have been inspirations to Jodi. Adina received invaluable support early on from her parents, Aliza and Fred Shima, Sylvain Merenlender and Francoise Calman; her academic advisors, David Woodruff, Andrew Dobson, and Chung-I Wu; and everyone at Stanford's Center for Conservation Biology. Today she is privileged to have fantastic colleagues at UC Berkeley and to have the support of the Integrated Hardwood Range Management Program. Her current students—Emily Heaton, Sarah Reed, Allison Bidlack, Juliet Christian-Smith, Adena Rissman, and Ted Grantham—are champions for their hard work and patience. Adina worked on this book while on sabbatical at the University of Queensland, and she is grateful to Karen Richardson, Hugh Possingham, and everyone in the Spatial Ecology Lab for many incredible experiences.

We also want to thank the many folks who contributed directly to this book, providing advice, information, figures, or review of chapters. These include Alexei Andreev, Misti Arias, Luciano Bani, Reginald H. Barrett, Jon Beckmann, Paul Beier, Steve Beissinger, G. S. (Hans) Bekker, Joel Berger, Hein D. van Bohemen, Douglas Bolger, Colin Brooks, Whisper Camel, Sue Carnevale, Cheryl Chetkiewicz, Juliet Christian-Smith, Tony Clevenger, Charlie Cooke, Caitlin Cornwall, Tim Crawford, Richard

Dale, Brent J. Danielson, Heather Dempsey, Joe DiDonato, Brock Dolman, Tim Duane, Sally Fairfax, Doris Fischer, Charles Foley, Lara Foley, Karen Gaffney, Joshua Ginsberg, Dennis Glick, Craig Groves, Amanda Hardy, John Harte, Kerry Heise, David H. House, Marcel Huijser, Bob and Kris Inman, Karen Klitz, Claire Kremen, Lorraine LaPlante, William F. Laurance, Raymond-Pierre Lebeau, Domingos Macedo, Andrea Mackenzie, Dale McCullough, Seonaid Melville, David Moffet, Katia Morgado, Karl Musser, Jeff Opperman, Richard S. Ostfeld, Justina C. Ray, Kent H. Redford, Karen Richardson, Adena Rissman, Eric Sanderson, Graca Saraiva, Jessica Shepherd, Peter B. Stacey, Robert C. Stebbins, Paul Sutton, Gary Tabor, Severine Vuilleumier, Bill Weber, Kerrie Wilson, and Steve Zack.

We also want to thank Brent Brock, Amber Shrum, Jane Rohrbough, and Andra Toivola for their able assistance in preparing art work and other aspects of manuscript production. Karen Klitz prepared figures for Chapters 3 and 5. Jeff Burrell spent significant time on art work and reference formatting.

We much appreciate the resources provided by the University of California at Berkeley, especially the awesome Biosciences Library. The Wildlife Conservation Society, and in particular Bill Weber and Eva Fearn, supported Jodi in allocating time for this project and hosted a critical meeting for the three of us in Bozeman, Montana. At Island Press, editors Barbara Dean, Laura Carrithers, and Bridhid Willson provided important feedback and encouragement throughout the process.

Last and most important, we thank our immediate families for support, especially given the many weekend and evening hours needed to complete this project. These include Jodi's husband and best friend, David House, and Jesse Hilty House, who accompanied the process both in the womb and then in a sling in her early days of life, perhaps the initiation of a future conservationist. Louise Lidicker was generous in her support and patience throughout this major undertaking. Kerry Heise, Adina's husband, was extremely supportive and a fantastic father. Noah Douglas (seven) showed great interest in Mama's book, and Ariella Rachel (five) gave up too many hugs.

Introduction

When Christy Vreeland worked at the Sonoma Development Center, a public facility in Sonoma Valley, California, that serves the developmentally disabled, she began to take note of the oak woodlands around her. She realized that some of the Center's lands and some adjacent properties were among the few places in the valley floor that were, as yet, relatively undeveloped. Moreover, she noticed that most of the protected areas in the region were isolated on surrounding mountaintops. So in 1995, she walked into the Sonoma Ecology Center, a local watershed conservation organization, with a plan to conserve a wildlife corridor across the valley floor. She was determined to protect what she referred to as the last dark area in the valley at night—a corridor of natural vegetation joining two adjacent mountain ranges that she hoped animals could move through.

The year after Christy approached the Sonoma Ecology Center with her plan, the three of us began looking for study sites to conduct research on wildlife use of riparian corridors in Sonoma County for Jodi's dissertation research. We heard of Christy's plan through environmental contacts and got in touch with her, and she gave Jodi a tour of the valley, including the proposed habitat corridor, which stands out among vineyards and residential development. Jodi's subsequent research, as well as a steady stream of reported wildlife sightings, indicated that Christy had indeed identified an area in the valley that many wildlife species used. Christy passed away in 2003 after a long fight with cancer, but her vision lives on today and is helping keep wildlife in Sonoma Valley. The Sonoma Ecology Center continues to pursue the protection of that habitat corridor, which is discussed in Chapter 9.

As in Sonoma Valley, landscapes across the world are often a mix of urban and rural development, agriculture, and remnant patches of natural habitat. With uninhabitable land and protected areas providing the only barriers to further development, the human footprint continues to expand, resulting in habitat fragmentation and isolation of habitat patches. This in turn reduces opportunities for the movement of organisms among the patches. Ecological, social, and economic problems often arise as a result of fragmentation caused by humans.

To confront the ecological problems associated with habitat fragmentation, conservation science has focused on protecting and enhancing connectivity by maintaining and restoring landscape linkages in the form of corridors. This book focuses on connectivity as a measure of the extent to which plants and animals can move between habitat patches. We refer to landscape features such as corridors, greenbelts, and ecological networks as potential means for achieving connectivity.

Extensive private and public effort is expended on protecting and restoring natural or seminatural passageways to maintain and enhance connectivity. Actions vary in purpose and scale from building highway underpasses to reduce amphibian road kill to linking vast continental ranges for migrating herds of ungulates. Yet the utility and effectiveness of these efforts for increasing species' persistence remain relatively unknown.

To increase our understanding of how species are impacted by fragmentation, investigators address different aspects of connectivity. Topics of research include the effects of distance between fragments, habitat quality both in remaining patches and in land and water surrounding these patches, and how variations in species' life history attributes affect species movement between fragments. Likewise, researchers have begun to explore how dispersal of organisms, home range dynamics, and migration are affected by habitat discontinuities. Some efforts focus on single-species movement, while others examine ways of maintaining community coherence. Some applied research addresses natural habitat connectivity by examining the suitability of corridors as compared to adjacent land-use types for movement by animal and plant species.

Despite existing research on these issues, there has been surprisingly little interpretation of the research results into practical guidelines for retaining connectivity. Lack of strong guidelines from the scientific community has not inhibited conservationists and land managers from wholeheartedly adopting the corridor concept, however. Some conservation organizations have been formed explicitly to address connectivity as a way of conserving biodiversity. Other organizations interested in conserving or restoring land consider connectivity in setting priorities for their expenditures. These practical conservation decisions are often made with little reference to the scientific findings pertaining to connectivity and perhaps could be strengthened by extracting lessons learned from research and prior conservation projects. In an effort to address the gap between science and conservation practice, this book examines the science behind the common assumptions about habitat connectivity and the utility of corridors.

Cities sometimes choose to preserve greenways in an effort to improve the quality of human life by limiting urban sprawl and providing opportunities for aesthetic enhancements and recreation. However, greenways may also serve as de facto connectors and habitat for biodiversity. Unfortunately, planning efforts rarely account for the requirements of human and natural systems to coexist and rarely consider potential conflicts or synergisms between the two. This leaves individuals and organizations attempting to protect and enhance connectivity without guidelines that incorporate the relevant human context in which corridors must function. In this book we review and assess the scientific concepts and evidence that lead to the use of corridors as a conservation strategy. Then we provide guiding principles as well as cautionary notes for those implementing corridor projects. Part I begins by documenting the need for such guidelines and the scientific theories that support the current conceptual and practical directions of this field. In the first chapter we set the context of land-use change that leads to the need for restoring connectivity. Chapter 2 reviews the consequences of habitat fragmentation for biodiversity conservation. The ecological principles that constitute the scientific underpinnings for our understanding of the pattern, process, and consequences of fragmentation are covered in Chapter 3.

Part II details the methods used to achieve connectivity, and its potential benefits, as well as the pitfalls for biodiversity conservation. Chapter 4 surveys the various approaches to facilitating connectivity across the landscape, including greenbelts, wildlife passageways, hedgerows, and others. The benefits of these types of landscape features for wildlife and people are discussed. The discussion in Chapter 5 is devoted to the importance of the landscape surrounding core habitat areas, often referred to as the *matrix* by conservation biologists. It is widely assumed that there are environmental benefits of corridors and similar landscape features, but the evidence for that is sometimes equivocal. In Chapter 6 we critically review the diverse problems that can be associated with relying on corridors and other linkages for connectivity.

The final part of this book provides information on corridor design, methods to improve local and regional planning, and ways to conserve habitat links over the long term. Chapter 7 focuses on how different types of habitat and species requirements influence the type and configuration of desirable corridors. Chapter 8 emphasizes systematic conservation planning and reviews mapping and analysis tools that inform decisions related to increasing connectivity across the landscape. These include land cover mapping, land-use change analysis, prioritizing site selection, and examining the relative benefits of various design scenarios. The final chapter discusses considerations needed for implementing a corridor project based on applied examples. The brief conclusion reflects on what we have tried to accomplish with this book and suggests possible future research directions.

Our goal is to provide guidelines that combine conservation science and practical experience for maintaining, enhancing, and creating connectivity among areas of remaining natural habitat for biodiversity conservation. This book will serve scientists, landscape architects, planners, land managers, decision makers, and all those working on land and wildlife conservation. We attempt to engage a readership from a variety of disciplines including the social, biological, and physical sciences, a spectrum that encompasses those concerned with both natural and human-dominated systems. We are especially anxious to promote appreciation for the importance of connecting humans and natural systems in land planning efforts.

PART I

Why Maintain and Restore Connectivity

Expansion of human land use has resulted in widespread loss and fragmentation of natural habitat, which could lead to the largest global extinction event in history. Corridors, routes that facilitate movement of organisms
between habitat fragments, are increasingly being adopted as a tool to
maintain and restore biodiversity. We begin Part I of this book with background on land-use change, the primary cause of habitat loss and fragmentation. Next, we discuss the consequences of habitat fragmentation
for biodiversity, and, finally, we provide an overview of ecological theories
that are fundamental to studying fragmentation and connectivity. Hence,
Part I documents the need for increasing connectivity that is fueling the
corridor concept, as well as giving the critical ecological concepts that we
need for effective responses.

In Chapter 1, we explore the primary reasons for land-use change,
quantify the associated global transformation, and demonstrate how the
conversion of natural habitat is affecting natural and human-modified systems. In addition, we emphasize the limitations of relying entirely on protected areas for biodiversity conservation. We also discuss the importance
of private land conservation and some widely used regulatory and incentive-based tools. Most of the land on earth has been converted for activities such as urban and residential development, forestry, intensive
agriculture, and livestock grazing. As one of the primary impacts of land-
use change is habitat fragmentation, the consequences of fragmentation
for biodiversity conservation are addressed in Chapter 2. There we explore
the impacts of fragmentation on species and communities; smaller and
more isolated habitat patches alter community composition and threaten

the persistence of many species. We also compare natural and human-induced fragmentation and discuss the particular consequences of the latter. Chapter 3 explains the science that underlies the development of corridor conservation. We describe the scientific principles and theories that scientists and conservationists draw on to design and implement landscape linkages, including island biogeography, metapopulation theory, and metacommunity theory. Without addressing the need for such linkages, it will be difficult to maintain native species in increasingly human-dominated landscapes.

CHAPTER 1

Background: Land-Use Change and Habitat Loss

Habitat loss and associated species loss are primarily a result of the acceleration of land-use changes begun over the past century. Therefore, it is important to study land-use change as the root cause of the biodiversity crises. The human population has increased sixfold since the 1800s, and the earth has been transformed to accommodate human habitation and rising consumption (Ehrlich and Ehrlich 2004). A human footprint is detectable across 83 percent of the land area in the world, excluding Antarctica (Sanderson et al. 2002; Figure 1.1). Land-use change associated with human development represents one of the most serious threats to terrestrial biodiversity, along with climate change, nitrogen deposition, and invasive species (Sala et al. 2000).

Cultivation of North America's Great Plains (2.6 million square kilometers [1 million square miles]), for example, has led to the loss of more than 96 percent of its tallgrass prairie (Samson and Knopf 1994), 76 to 82 percent of its eastern mixed grasslands, and 25 percent of its shortgrass prairie (Loveland and Hutcheson 1995). The habitat loss from such land-use change due to agricultural expansion has led to a corresponding loss of biodiversity. In the Great Plains, 465 species are of conservation concern, and endemic songbird and grassland nesting bird species have declined by 50 and 75 percent, respectively (Environmental Protection Agency 2002, Ostlie et al. 1997).

An early attempt to map large undeveloped wilderness areas (greater than 4,000 square kilometers, or about 1,500 square miles) globally estimated that such areas constitute only 16 percent of the land on earth outside the polar regions (McCloskey and Spalding 1989). The paucity of

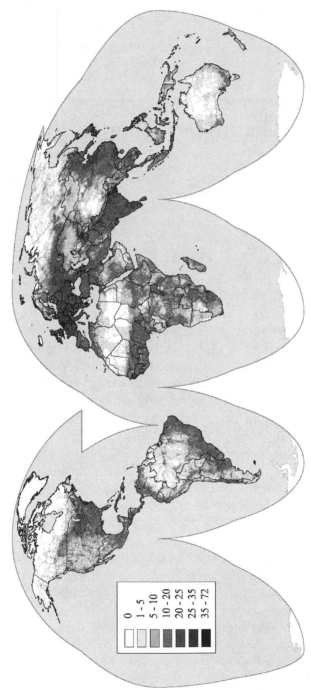

Figure 1.1. Impact of the human footprint worldwide. The darker the area, the greater the impact, based on a relative numerical score. The mapped relative impact scores are derived from the following factors: human population density greater than one person per square kilometer (0.39 square mile); agricultural land use; built-up areas or settlements; access within 15 kilometers (9 miles) of a road, major river, or coastline; and nighttime light bright enough to be detected by satellite sensors. (Permission to use by Eric Sanderson, Wildlife Conservation Society and Center for International Earth Science Information Network at Columbia University).

0
1 - 5
5 - 10
10 - 20
20 - 25
25 - 35
35 - 72

wilderness on earth restricts the space suitable for persistence of some species, especially those that require large ranges. Where wilderness does remain it is often in isolated patches. These isolated patches of natural habitat rarely contain the biodiversity that existed in the region prior to fragmentation. For example, in the Amazon basin, researchers discovered that because of the way species are distributed across the landscape, some species will not be supported in remaining forest fragments (Laurance et al. 2002).

Synergistic effects between habitat loss, habitat fragmentation, and global climate warming (Opdam and Wascher 2004) can compound the effects of habitat loss on biodiversity. For example, researchers are trying to determine whether species will be able to shift their distributions or evolve new adaptations fast enough to accommodate global climate change. Numerous studies document the historical movement of species as climate changed in the past (DeChaine and Martin 2004), but the unprecedented speed of modern climate change combined with habitat loss could make historical processes of adaptation less applicable today. Habitat fragmentation and global warming may not generally present a significant threat to species that currently survive in a diverse array of habitat patches, but species that are isolated in only a few patches or restricted to mountaintops may not be able to rapidly shift their distribution in order to survive climate change (Opdam and Wascher 2004).

Widespread habitat loss and fragmentation due to human activities clearly threatens species survival (Kucera and Barrett 1995). The global rate of species extinction is orders of magnitude higher today than before modern rates of land and water expropriation by human beings. In fact, if the current rate of biodiversity loss continues, we will experience the most extreme extinction event in the past 65 million years (E. O. Wilson 1988). These losses will be devastating for humans, given that we depend on the goods and services that intact ecosystems offer. Many of our medicines and all of our food and fiber, the basis for our economies and our survival, are derived from wild species. Ecosystems are also responsible for many of the natural processes on which we depend, such as maintaining air quality, soil production, and nutrient cycling; moderating climate; providing fresh water, fish and game, and pollination services; breaking down pollutants and waste; and controlling parasites and diseases (Naeem et al. 1994). An interdisciplinary field of study has arisen around trying to understand *ecological resilience*, which is defined as the capacity of a system to withstand changes to the processes that control its structures (Holling

and Gunderson 2002). One of the primary observations is that disturbing ecosystems can reduce their resilience and result in dramatic shifts to
less desirable states that weaken their capacity to provide ecosystem goods
and services (Folke et al. 2004).

Some scientists have quantified ecosystem services in financial terms
(Daily and Ellison 2002). One study estimated that the earth's biosphere
provides US$16 to $54 trillion worth of services per year that we currently
do not pay for (Costanza et al. 1997). While quantifying ecosystem services may help enlighten people to the importance of natural systems in
their daily life, it is difficult to quantify the value to humans of each of the
10 to 30 million or so species inhabiting the earth. The ecological roles
of most species are unknown to us, but we do know the key roles that
some species play in the normal functioning of biotic communities. For
example, if it were not for a few kinds of microorganisms that can digest
chiton, the shed exoskeletons of arthropods would bury the surface of the
earth. A study of the functional role of species in ecological communities
reveals that while multiple species can contribute to ecosystem process and
function in the same way, referred to as *redundancy* (B. H. Walker 1992),
ecosystem reliability may depend on that redundancy. In other words, removing species because they are redundant could have consequences
(Naeem 1998).

While the cost of species extinction to ecosystems and the goods and
services that they provide should be one of today's primary concerns, what
motivates many people to conserve nature is the intrinsic value of biodiversity (Rolston 1988). With the loss of biodiversity we are losing important opportunities for personal inspiration and cultural enrichment, whether
by bird watching, catching and releasing wild salmon, or simply enjoying a
natural scenic view. To ignore the emotional, and for some spiritual, importance of conserving biodiversity and focus solely on the goods and services people rely on is a mistake. We must acknowledge our ethical and
moral responsibility to prevent irreversible change to the earth's systems
so that we do not harm other species and our own future generations.

Importance and Types of Land-Use Change

Land-use change has received less public attention than other threats to
natural systems, such as global climate change and air and water pollution. This is true despite the fact that it is the primary cause of habitat loss
and ecosystem degradation and greatly exacerbates most of the other

threats to the environment (Harte 2001). For example, the impacts of land-use change on local climate can often be greater than predicted impacts of climate change because changing land cover alters the way the sunlight is absorbed and affects evapotranspiration rates (Lashof et al. 1997). In the Amazon basin, upland land-use change and associated forest loss have resulted in less local rainfall because less water is evaporating into the atmosphere from the rain forest (McGuffie and Henderson-Sellers 2004).

Land-use change has multiple drivers and complex outcomes that generally elude our ability to provide a technical fix for the problems generated. We cannot usually find a single solution to slowing land-use change or reducing its impacts that is as tangible as lowering carbon dioxide emissions to slow global warming, for example.

Given the interactions between human development patterns and the natural systems that support ecological processes and environmental goods and services, it is important that we understand the rates and patterns of land-use change resulting from human development (Tjallingii 2000). Addressing the process and pattern of past land-use change, forecasting future land-use change, and determining the risk of those changes to the environment are important for planning a sustainable future. The first step is to establish a causal relationship between land use and environmental impacts. We do not know the full impacts of land use on biodiversity over time, in part because we do not fully understand ecosystem resilience, the amount of disturbance that an ecosystem can withstand without changing its self-organizing processes and the variables that control its structures (Gunderson et al. 2002).

Land-use changes and the resulting impacts to natural habitats often happen incrementally with seemingly subtle effects. It can be difficult to assess the cumulative effects of those incremental changes, and most environmental impact assessments do not take them into account because analysis is done at the site scale and does not evaluate the potential for larger-scale impacts. This makes it hard to stop land-use change, even if scientists could quantify when thresholds of irreparable damage are crossed. And stopping destructive land use today because of the cumulative impacts of past activities places the entire cost of preventing development or mitigation on a few current operators, which can be politically, ethically, and even legally untenable.

The types of land-use change responsible for converting most natural areas around the world include agriculture, logging, and to some extent mining, and urban and residential development.

Agriculture, Logging, and Mining

The pressure during the Green Revolution, the period of technological development of agricultural practices that began in the 1940s, to increase cereal crop yields through the use of improved crop varieties, fertilizers, pesticides, irrigation, and mechanization led to more intensive agricultural production. Intensive agriculture was encouraged by subsidies and other governmental support measures, such as the Common Agricultural Policy, which established rules and regulations for Europe's agricultural sector beginning in the 1960s. Subsidies were put in place that rewarded farmers according to the volume of crops they produced, creating a strong incentive for high-input, high-output agriculture. As a result, the structure and species composition of agroecosystems were greatly simplified through a reduction in the types of crops grown and the removal of native vegetation in and around the farm to increase production. For example, there is some evidence that this led to declines of farmland birds in Europe (Donald et al. 2001).

In the past fifty years, a widespread movement toward sustainable agriculture has arisen as a backlash against the practices of the Green Revolution. Sustainability rests on the principle that we must meet today's agricultural and economic needs without degrading the natural resources for our future welfare. Sustainable agriculture and the study of agroecology remain primarily focused on practices and systems at the scale of the individual farm. Reforestation, planting hedgerows of native species, reducing the application of fertilizers and pesticides, and maintaining and restoring riparian corridors are treatments employed as part of agri-environment measures.

Some agri-environment programs provide financial incentives for farmers who implement environmentally friendly practices. In Europe, over 20 percent of the agricultural land today is under such programs. These types of incentive programs to restore farmland for environmental benefits are also receiving support in Australia under the Commonwealth's National Action Plan for Salinity and Water Quality (2000) and in the United States through the Department of Agriculture Farm Bill (2002). Also, in 1997, as part of the Conservation Reserve Program, the U.S. Department of Agriculture's Natural Resource Conservation Service developed a "buffer initiative," which promotes the use of contour buffer strips, field borders, grassed waterways, filter strips, and riparian forest buffers across America's farmland. These treatments represent various ways of planting around the farm to increase beneficial insects, filter sediment, slow runoff, and protect water courses. In some case, these linear landscape features can increase habitat connectivity. Recent research in Holland comparing the extent to which fields with and without these features support plant, bird, fly, and bee diversity revealed some benefit to invertebrates but none for plants and birds (Kleijn et al. 2004). There is a prevailing view that these landscape features will facilitate the movement of wildlife between natural areas and increase interbreeding between isolated populations, but that remains relatively untested.

Agriculture is expanding in approximately 70 percent of the countries of the world, and in two thirds of those countries forest area is decreasing (Food and Agriculture Organization of the United Nations [FAO] 2003). While agriculture can clearly result in deforestation, it usually receives relatively little environmental regulatory oversight compared to forestry and commercial and residential development (Giusti and Merenlender 2002, Hobson et al. 2002). This means that environmental impacts such as habitat loss and fragmentation resulting from agricultural expansion are rarely reviewed prior to agricultural land conversion.

The concept of sustainable agriculture needs to go beyond the boundaries of the farm to include larger ecosystems, the myriad of species those systems support, and the goods and services they provide to humans. Enhancing large-scale agricultural landscapes or "farmscapes" for biodiversity conservation represents a new approach to agroecology that recognizes the importance of sustainable agriculture at a landscape scale (Kremen et al. 2002). This approach goes beyond planting and retaining hedgerows and other habitat enhancements and addresses how farmland and wildlands influence one another. For example, researchers demonstrated enhanced pollinator activity in coffee plantations near remnant tropical forest in Costa Rica (Ricketts 2004). The importance of mitigating habitat fragmentation across the larger agricultural landscape has recently been recognized in Europe. To that end European environmental and agricultural policies now address connectivity with ecological networks (Jongman 1995). The establishment of ecological networks across Europe includes identifying core areas, corridors, buffer zones, and areas where restoration is needed to enhance those elements of the network (Jongman 1995).

Logging also continues to be a primary cause of deforestation, with 5.6 million hectares (13.8 million acres) logged globally each year from 1981 to 1990 (FAO 2003). In tropical countries, deforestation is fragmenting the landscape, resulting in an increased number of remnant forest patches within a human-modified matrix (Bierregaard et al. 1992). In some places, especially in the northern latitude locations that were logged more extensively in the past than they are now, forest cover is increasing due to forest regeneration (FAO 2003). In most countries, forestry is regulated, in part for environmental protection. These regulations often include the retention of corridors, especially along water courses, but few generalizations can be made about the design and ultimate utility of these efforts (Bierregaard et al. 1997, Simberloff 2001). Despite some efforts by industry to practice sustainable forestry and maintain connectivity, logging is still one of the primary causes of land-use change, leading to habitat loss and fragmentation in many parts of the world.

Mining and energy resource development can severely affect biodiversity locally and are becoming more of a problem, as well, especially in developing countries. With the assistance of foreign direct investment, mining has increased in sub-Saharan Africa. In Ghana, for example, foreign direct investment primarily for gold extraction increased rapidly from US$15 million in 1990 to US$233 million in 1994 (Boocock 2002). Changes in land use as a result of gold mining in Ghana include loss of farmland and loss of natural habitat within forest reserves (Boocock 2002). The problems of mining activities are well documented and include water and soil contamination, which cause severe health problems for local people (Hilson 2002). Surrounding flora and fauna are, of course, affected, but less appreciated are the diversion of rivers to provide water for mining activities and the increase in roads, both of which have regional environmental consequences.

Residential Development

Urban expansion is the most well-studied and modeled type of land-use change worldwide. The prevailing view of urban development in the early 1900s was of concentric circles of land use starting with the densest development and radiating out to lower-density housing development (Burgess 1924). After World War II, development was no longer contained within cities, as the movement to the suburbs became widespread in the United States and elsewhere. In the United States, this resulted in a doubling of high-density developed areas from 1960 to 1990 (Theobald 2001). In some European countries, the number of suburban residents has doubled or tripled over the past fifty years, while the number of people living within large cities has slightly decreased in the same time period.

On the other hand, a tremendous amount of recent urban development has occurred in China, and the urban-rural boundary is eroding there (M. Y. L. Wang 1997). In an attempt to maintain the rural character surrounding urban areas and to prevent sprawling development outside of cities, the conservation of greenbelts has become popular in many places. Greenbelts are linear open spaces that are often created along roadways and rivers and are designed to prevent cities from merging and to provide visual and sometimes recreational amenities to the public. The utility of this approach is not well documented, however. Urban expansion can often leapfrog over greenbelts and eventually surround them, turning them into islands in a sea of development. Nevertheless, greenbelts and greenways may offer some conservation benefits, depending on design, activities permitted, and landscape context (Chapter 4).

Containing urban development through greenbelts and urban growth boundaries, which set the maximum area to be developed by restricting water and sewer services, does not necessarily reduce conversion of rural areas, which are rapidly being impacted by exurban development. Exurbia, low-density residential development outside of urban growth boundaries (Figure 1.2), is now the fastest-growing type of land use in the United States (Theobald 2001). It is different than the dense suburban development that commonly occurred from 1960 to 1990. Exurban development results in an unorganized scattering of homes on large parcels of land (1 unit per 4–16 hectares or 10–40 acres) along rural roads that do not have streetlights and rely on private water wells and individual sewage systems (Theobald 2001). Such residential development is popular in most of the developed world today.

Despite the potential for enormous impacts to wildlands, this type of land-use change has received little attention from the scientific and environmental conservation communities. The increased rate of exurban development, along with the larger land area required to support it, means that ten times the amount of land in the United States was converted to low-density development compared to urban densities in 2000 (Theobald 2001). Estimates based on nighttime satellite imagery suggest that 37 percent of the U.S. population now lives in exurban areas that account for 14 percent of the land area (Figure 1.3). Purely urban areas, including traditional dense suburban development, account for only 1.7 percent of the land area and house 55 percent of the population. Rural areas (84 percent of the land area) contain only 8 percent of the population (Sutton et al. 2006).

Figure 1.2. Exurban residential development on the west side of the Bridger Mountains north of Bozeman, Montana over the last decade. (Photo by Tim Crawford.)

The extent of exurban development appears to be increasing across much of the developed world. For example, between 1971 and 2001, Alberta experienced the greatest increase in rural population in Canada, 32 percent, compared to the overall growth of 8 percent recorded for the rural Canadian West (Azmier and Stone 2003). Country homes are common across Europe, and especially in France, where an explosion of country home building started in the 1970s (Dubost 1998). The number of rural residents has also increased dramatically in Denmark (Tress and Tress 2001). In the Netherlands, large estate homes are very popular, and attempts are being made to make these "new rural lifestyle estates" pay for restoring agricultural land for conservation purposes (van den Berg and Wintjes 2000). What started off as simple country outposts in Russia (dachas) are now year-round homes in the exurbs of Russian cities (Struyk and Angelici 1996). In Hungary, and particularly outside Budapest, issues surrounding the rate and extent of sprawl have received some attention (Éri 2001). Yet, despite the widespread nature of this type of land use and its rapid rate of spread, little is known about the processes that drive exurbia or its environmental consequences (Box 1.1, Figure 1.4).

Figure 1.3. Exurban development, indicated by the gray areas surrounding black urban centers, across the United States. Mapped by Paul Sutton, who identified those areas as having low light levels from nighttime satellite imagery and forest cover based on Landsat satellite imagery (Sutton et al. 2006; permission to use by Paul Sutton).

It is known that the patchwork of land use that results from exurban development can create social and economic problems for human systems. The spread of rural residential development can cause farmland fragmentation and result in social and economic impacts that are generally under-appreciated. Agricultural landscapes of sufficient size and connectivity are essential for the production of food and fiber and the maintenance of some rural cultures. Sometimes the level of production necessary to support important collective infrastructure in a region, such as processing, packing, and transportation facilities, is threatened when farmland is taken out of production. Conflicts between farmers and their rural residential neighbors can arise over noise, chemical applications, and odors that are part of farming (Kay et al. 2003).

There are health risks, social stresses, and increased costs associated with increased commuting distances for those living in rural areas. For example, there is an increased fire threat in the urban-wildland interface due to exurban development, as demonstrated through analysis of the regional fire

Box 1.1.

Consequences of Exurban Development for Bird Species

Even at low densities, human development can impact native species. Disturbances associated with exurban development include roads, domestic pets, and great densities of urban-adapted predators such as raccoons and rats, as well as nest parasites such as brown-headed cowbirds. Since exurban expansion is a fairly recent phenomenon, there have been relatively few attempts to quantify its impacts on biodiversity. However, a few papers do include field studies of birds in exurbia (Maestas et al. 2003, Merenlender et al. 1998, Odell and Knight 2001, Odell et al. 2003). For example, a study in one Colorado watershed by Maestas et al. (2003) found six species at higher densities in exurban sites than on ranches and reserves: black-billed magpie (*Pica pica*), European starling (*Sturnus vulgaris*), Brewer's blackbird (*Euphagus cyanocephalus*), American goldfinch (*Carduelis tristis*), house wren (*Troglodytes aedon*), and broad-tailed hummingbird (*Selasphorus platycercus*); whereas green-tailed towhee (*Pipilo chlorurus*), vesper sparrow (*Pooecetes gramineus*), and rock wren (*Salpinctes obsoletus*) were less abundant in exurban sites. Also, many more domestic dogs and cats were present at exurban sites.

To explicitly address the influence of an exurban home on the surrounding bird community, Odell and Knight (2001) surveyed songbirds at three different distances from forty homes. They found significant increases in the density of six bird species 30 meters (98 feet) from the homes, compared to 180 meters (590 feet) and 330 meters (1,083 feet). The six species were American robin (*Turdus migratorius*), brown-headed cowbird (*Moiothrus ater*), house wren, black-billed magpie, Brewer's blackbird, and mountain bluebird (*Sialia currucoides*). The same birds were also more common in high-density housing developments (1.04 ± 0.67 houses per hectare) compared to lower-density developments (0.095 ± 0.083 houses per hectare) and undeveloped sites (at least 700 meters, or 2,297 feet, from any development), with the exception of Brewer's blackbird and the European starling. In the natural habitat, farther from exurban home sites, the following species had significantly higher densities compared to surveys 30 meters from the homes: black-capped chickadee (*Parus atricapillus*), black-headed grosbeak (*Pheucticus melanocephalus*), orange-crowned warbler (*Ver-*

mivora celata), blue-gray gnatcatcher (*Polioptila caerulea*), dusky fly-catcher (*Empidonax oberholseri*), and spotted towhee (*Pipilo maculates*). The same species, with the exception of the spotted towhee and the addition of the green-tailed towhee and the plumbeous vireo (*Vireo plumbeous*), were also at higher densities in undeveloped sites than in higher-density housing areas. These studies demonstrate that some urban adapter species are found close to exurban homes but not necessarily in the surrounding habitat.

A similar study was conducted in California (Merenlender et al. 1998) comparing suburbs (one house per 0.1–1 hectare, or 0.2–2.47 acres), exurban areas (one house per 4–16 hectares, or 10–40 acres), and undeveloped landscapes (large undeveloped contiguous areas greater than 400 hectares, or 1,000 acres, in ownerships greater than 120 hectares, or 296 acres). More recent unpublished data from these sites revealed significant differences in the detection rates for Hutton's vireo (*Vireo huttoni*), orange-crowned warbler, and northern flicker (*Colaptes auratus*) (Figure 1.4) at undeveloped sites

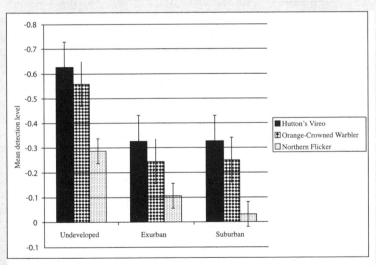

Figure 1.4. Detection levels of Hutton's vireo, orange-crowned warbler, and northern flicker in undeveloped, exurban, and suburban sites. All three were found significantly more often in undeveloped sites compared to exurban and suburban sites, where they were detected at similar levels. These data come from ninety-six point count locations in Sonoma County, California, that were evenly divided among the three levels of residential development.

compared to exurban and suburban areas. For these species, the exurban sites had similar detection rates as the suburban sites. The following species were more common in the suburban sites: California quail (*Callipepla californica*), oak titmouse (*Baeolophus inornatus*), house finch (*Carpodacus mexicanus*), California towhee (*Pipilo crissalis*), northern mockingbird (*Mimus polyglottos*), American crow (*Corvus brachyrhynchos*), and turkey vulture (*Cathartes aura*). This suggests that urban adapters are more common in suburbs than in exurbia; however, more sensitive bird species may be impacted by low-density exurban development to the same extent as observed in traditional suburbs.

regime and the increased need for evacuation plans (Wells et al. 2004, Cova and Johnson 2002). More surprisingly, in the United States, combined rates of traffic fatalities were higher in some or all exurban areas than in central cities or inner suburbs (Lucy 2003). Exurbia requires more road per home, and people who live there drive more, exacting a greater cost on the public and the environment. For example, in Jefferson County, Colorado, eighteen houses are found per mile of road on average compared to fifty-nine in cities (Best 2005). More roads, carbon dioxide, septic systems, wells, and increased fire protection are just a few of the underappreciated consequences of exurban expansion, and the associated costs fall on the general public (Best 2005).

Working Landscapes

Areas susceptible to exurban development are often "working landscapes," predominantly under extensive agricultural use (forestry and livestock grazing). Extensive agricultural use has resulted in landscape-level changes in many parts of the world. Clearing for rangeland has been a significant driver of historic forest loss and in some areas is still occurring, presenting a problem in many places today. For example, much of the deforestation in the Brazilian Amazon is attributed to cattle ranching (Fearnside 1993). These changes accelerated in the early 1900s, in part due to mechanization, road development, and new global markets. Rangelands represent a significant amount of open space in many countries. For example, 70 percent of Australia is covered by semiarid rangelands, and in the United States 109 million hectares (270 million acres) of federal public land is grazed rangeland in eleven western states.

The economic viability of open-range grazing operations is marginal in most parts of the world today, making the reduction and removal of livestock from the range commonplace. Cattle are being removed from the land for economic reasons or, in some cases, through government incentive programs designed to help conserve natural resources. In the United States, most cattle production is now consolidated in feedlots. Between 1982 and 2002, nonfederal acreage devoted to grazing uses declined from 247 million hectares (611 million acres) to 234 million hectares (578 million acres), a decrease of over 5 percent, and much of that land has been developed (Natural Resource Conservation Service 2002).

Working forests and grazing lands provide extensive open space and habitat for a myriad of species. Such areas are the focus of conservation efforts in many parts of the world where the land in protected parks is not sufficient to sustain natural ecosystems. Protecting extensive agricultural land from more intensive development, such as housing and row crops, is a high priority for biodiversity conservation. Therefore, it is important that we consider how these open spaces will be protected from exurban development and other land-use change, given that activities such as livestock grazing and forestry may not be economically viable in the long term.

Conserving working landscapes that are threatened by more intensive development, including exurban development, may represent one of the most important yet difficult challenges that some places face with respect to conserving larger landscape connections for wildlife. There are few known solutions to the problem of exurban development, because landowners are simply exercising their right to subdivide their land into the minimum allowable parcel size. One option is trading development rights to remove such sites from the land supply for development in exchange for increased densities elsewhere. Another is to purchase development rights to prevent subdivision, a popular approach discussed further on and in Chapter 9 (Merenlender et al. 2004).

The Limitations of Protected Areas

While 11.5 percent of the land on earth is now designated as protected, those areas do not provide sufficient protection for global biodiversity (Rodrigues et al. 2004). The amount of habitat that has been converted exceeds the amount protected by a ratio of 8 to 1 in temperate grasslands and Mediterranean biomes, and 10 to 1 in more than 140 out of 810 ecoregions studied (Hoekstra et al. 2005). Retaining the planet's diversity of life forms will require additional conservation effort because protected areas are more likely to be designated in certain types of ecosystems and often on less productive land, and because biodiversity is not equally distributed across the globe.

When it comes to biodiversity, not all protected areas are equal. Some conservationists facetiously refer to national parks in the United States as "rock and ice" because of their placement in deserts and high mountains and the focus on conservation of unique geological features and landscape views (Weber 2004). There are, of course, exceptions such as Everglades National Park in Florida. Still, most public land in the United States is less productive, is less watered, and has lower-quality soils than private land (J. M. Scott et al. 2001). In regions such as the Greater Yellowstone Ecosystem (GYE), this pattern is clear, with most of the productive valleys falling under private ownership and the surrounding mountains in public ownership (Hansen and Rotella 2002).

It is not surprising that productive land has been converted for agricultural and other uses and is privately owned. But as a result, many species are imperiled. Over 90 percent of all federally threatened and endangered flora and fauna can be found on nonfederal land (Wilcove et al. 1996). The status of 51 percent of the species occurring only on private land is defined by the U.S. Fish and Wildlife Service as unknown, and extinction rates for poorly studied taxa are generally higher than for well-studied taxa (McKinney 2002). These statistics suggest that the future conversion of remaining natural habitat on private lands will impact species' persistence.

Most parks are too small to conserve resident species, in part because activities on adjacent private land can, in some cases, affect species' persistence. For example, Yellowstone National Park, the largest national park in the contiguous United States, is too small in and of itself to support viable populations of large carnivores for the long term (Newmark 1987). Some large carnivores, such as grizzly bear (*Ursus arctos*), cannot coexist on lands occupied even at low densities by humans because they prey on domestic animals and occasionally harm humans (Primm and Clark 1996). In recognition that Yellowstone National Park is inadequate for their long-term conservation, the GYE was delineated as the area necessary to support a long-term, viable population of grizzly bears (Craighead 1991; Figure 1.5). Between 1973 and 2001, 256 bear deaths were caused by humans across the GYE, whereas only 92 bears died from

Figure 1.5. The extent of the Greater Yellowstone Ecosystem (the larger area outlined in gray), which was delineated after biologists recognized that the existing Yellowstone and Grand Teton National Parks (outlined in black) would not adequately protect large carnivores and other wide-ranging animals.

other causes (Haroldson and Frey 2002). If human activities expand across the GYE landscape and the human population continues to grow as it has in recent years, increasing by 55 percent between 1970 and 1997, these kinds of conflicts are likely to increase (Hansen and Rotella 2002). Given the rising rates of human disturbance and declining habitat connectivity of the Greater Yellowstone Ecosystem to other large natural habitat areas, maintaining grizzly bears and other species with large area requirements in the Greater Yellowstone Ecosystem will become increasingly difficult.

The modified landscapes surrounding reserves can be critical for biodiversity conservation in the reserve, as in the case of the grizzly bear, and at the same time can negatively impact the protected areas. Problems from adjacent lands can include invasive species, poaching, pollution, and altered disturbance regimes such as flood and fire. In many cases, the natural processes required to maintain biodiversity operate at a larger scale than the reserve itself. For example, fire regimes, catastrophic storms, and top predators can operate on very large spatial scales and yet may be critical for the health and persistence of a given region's biota. The ecological integrity of the land beyond the reserves should be maintained to ensure the persistence of these types of large-scale natural phenomena.

Gar National Park in India is a classic example of a patch of habitat surrounded by very high human densities, and, in fact, it is regularly encroached on by human activities. It is also the only place in the world where Asiatic lion (*Leo leo persica*) still occur, although their numbers are very low due to limited habitat (Khan 1995, Oza 1983). Human and lion conflicts are an ongoing problem in this park. Because there are no other large islands of natural habitat in the nearby region and thus no opportunity to restore connectivity, the government has made a decision to translocate lions to another region of India as a means of helping the Asiatic lion (Cat Specialist Group 2000, Khan 1995).

Ultimately, those species that naturally occur at low densities and require large home ranges, such as Asiatic lions and many other predators, are potentially most impacted by habitat loss unless they can continue accessing sufficient natural patches to survive (Beier 1993, Mac Nally and Bennett 1997, Primm and Clark 1996). For example, in North America, a population of fifty to seventy mountain lions (*Puma concolor*) would need at minimum about 8,000 square kilometers (3,120 square miles), and a population of two hundred black bears (*Ursus americanus*) would need at least 2,000 square kilometers (780 square miles); both require landscapes larger than most protected areas (Harris et al. 1996). Such small populations of either species are probably not sustainable in the long term. Sustaining viable populations of these species requires much larger natural habitat areas that contain the food, den sites, and other characteristics needed for them to survive.

Clearly, protected areas alone will not provide sufficient habitat for wide-ranging species or sustain viable populations of all species. Maintaining and enhancing natural resources provided by privately or communally owned and managed landscapes are often the focus of efforts to reconnect our landscapes for biodiversity conservation (Kautz and Cox 2001). Therefore, issues related to private land and the conservation regulations pertaining to them influence our efforts to reconnect most fragmented landscapes. In some countries, legal pluralism is common, meaning more than one legal system (e.g., state or village customary law and religious law) or institution may jointly govern use of the land. Under those circumstances, the granting of land tenure to individuals can lead to ecological and social catastrophe (J. C. Scott 1999)—for example, when landowners make decisions based on individual short-term gain, as compared to sustaining collectively managed resources for future generations.

We should also not assume that public ownership is entirely separated from regional private interests. The history of how particular protected areas were implemented often influences the attitudes of people surrounding those sites, either negatively or positively. For example, the establishment of public lands has displaced some communities and impacted local economies. Such experiences can influence the way local people respond to managing for biodiversity conservation. In response to this problem and others, the responsibility of managing public land is being returned to local communities in some countries. So we see that differences between public and private lands and the consequences for resource conservation are not uniform and that many complex arrangements exist.

Environmental Regulation and Incentive-Based Conservation

Existing legal frameworks greatly influence the extent to which regulations can protect natural resources on privately owned land and, in turn, where incentives have to be provided. In the United States, the federal Endangered Species Act (ESA) is one of the most powerful environmental protection acts passed by Congress, because it makes the "taking" of listed species illegal. Any action deemed as a taking of an endangered species, including habitat destruction, can be prevented or punished by law. The Act allows any isolated population of a species to be listed for protection. It serves as a means to protect habitat, conserve individual species, and influence land use and management. In other countries, there are laws that protect diverse habitats explicitly, and environmental protection relies less on the protection of species.

In some situations, the ESA facilitates the establishment of habitat corridors as part of adopted Habitat Conservation Plans. These plans allow regional development to proceed, including some takings of endangered species. Efforts such as Natural Communities Conservation Plans (NCCP) and Multiple Species Conservation Plans (MSCP) protect existing endangered species in addition to protecting landowners from penalties for an allowable level of takings and from restrictions on land use associated with future potential species listings. Such habitat plans promote conservation planning among landowners to avoid the gradual loss and fragmentation of wildlands (McCaull 1994). However, their impact on habitat and species conservation remains to be demonstrated (Wilhere 2002). Identifying and protecting wildlife corridors are often an integral part of the NCCP and MSCP processes. For example, in southwestern San Diego, an MSCP preserve was designed to maintain connections between core habitat areas, including linkages between coastal lagoons and more inland habitats and between different watersheds (Conservation Biology Institute 2002).

The Endangered Species Act is a double-edged sword. It has provided strong legal protections against taking endangered species. Approximately 10 percent of the listed species are recovering, although we cannot say if that should be credited entirely to the Act. On the other hand, the legal powers to protect takings that can include habitat loss make most private landowners fear the consequences of harboring endangered species on their lands. A good example of this is the resistance of landowners in Potter Valley, California, to allow an endangered salmon migration corridor in the form of a bypass around an existing dam because they feared additional constraints on land use if endangered fish were to return to the creeks in the valley. Safe harbor agreements can be negotiated between nonfederal landowners and the federal government, however. These protect landowners from additional restrictions or prosecution if future incidental takings occur as a result of their conservation actions, including protecting and restoring wildlife corridors. Less litigious societies may not find it necessary to enter into these types of agreements with landowners interested in restoring habitat for endangered species.

The ESA takes precedence over all local and state laws and practices and all city and county planning and permits. But it is not the only law that can affect land use. In fact, most land-use decisions are made through regional plans and local zoning restrictions. Other environmental protection laws can also be important, such as the U.S. Clean Water Act. In some countries, habitat protection policies require the protection of a target amount of habitat, reducing the need to focus on threatened and endangered species as the only means of guaranteeing habitat protection (Soulé and Sanjayan 1998). Such policies are often implemented to meet the goals of international agreements, such as the 2010 targets outlined in the Convention on Biodiversity.

Local land-use regulations such as zoning for lower-density development, transfer of development rights, and urban growth limits are used to minimize loss of natural habitat and agricultural land. Environmental protection regulations also can limit or require mitigation for development that will damage the environment, as in the U.S. Clean Water Act. In some cases, the extraction of a natural resource is regulated, and such regulations often include environmental protections. That is the case with forestry regulations and mining laws. Regulations also exist for the sole purpose of protecting natural habitat, as with Australian legislation that bans land clearing in all states except Tasmania, where similar restrictions are pending.

The extent to which regulations limit the economic returns of landowners influences the level of protection that can be expected from such regulations without compensation. For example, in California, local land-use regulations sometimes restrict the clearing of native vegetation along stream courses. However, the width of stream buffers is usually determined by what is politically palatable without compensation, as well as the importance of these areas for resource conservation, rather than on the optimal width for environmental protection. The advantage of regulations is that they are applied at one time rather than piecemeal and are generally consistent across the jurisdiction. Another important advantage is that they reflect political decisions made by public representatives rather than private boards. In some cases, landowners are compensated for financial losses associated with changes to land-use regulations that occurred after they purchased the land. That is the case in Queensland, Australia, where funds are available to compensate landowners who were recently prohibited from clearing the bush. Also in Oregon, a law passed in 2004 provides for landowners to be compensated if land-use regulations diminish the value of their property by limiting development.

The use of incentive-based conservation arose in part as a backlash against the limitations caused by regulations for protecting private land resources (Merenlender et al. 2004). For example, conservation easements or covenants can, on a site-specific basis, apply enforceable land-use restrictions that supersede zoning and protect habitat from fragmentation and conversion to other uses. Typically, a conservation easement transfers some development and management options—such as the right to subdivide or to cut trees—from the fee holder to a nonprofit or government organization that holds those rights. The fee holder may sell or donate an easement while retaining certain rights, such as the right to build additional homes, add roads, or plant row crops. The fee holder pays property taxes only on the remaining value of the land and continues to own and manage the property within the bounds set by the easement. The easement holder is responsible for monitoring and enforcement of easement specifications. Conservation easements are a tool commonly used to increase connectivity among public reserves without incurring the costs and responsibilities of full ownership (see Chapter 9).

Reconnecting Our Landscapes

Our remnant protected areas are often too small to allow for the persistence of viable populations of species. Connecting networks of protected areas may increase species' persistence. The need to recover endangered species and rare habitat types has driven much of the demand for habitat connectivity. The most common solution is to maintain and restore habitat that will provide connections between protected natural areas for wildlife movement. However, it is not always clear that connecting wildlands through linear habitat features across a disturbed landscape enhances species' persistence within reserves. In many cases it is premature to suggest methods for enhancing connectivity when not enough is known about the requirements of focal species and whether increased connectivity will result in boosting their persistence. However, we recognize the need to move forward with plans to increase connectivity through corridors, ecological networks, and other landscape features to minimize continued fragmentation and associated species extinctions.

Many people who work to restore ecosystems and recover species have an understanding of the importance of biodiversity and a personal connection with their local flora and fauna. These provide them with important knowledge about nature and in some cases connections with past generations who once may have subsisted primarily on natural resources from the immediate area. For example, in the book *Totem Salmon*, Freeman House (1999) reveals the diverse reasons members of the Matole River watershed group in California worked together to restore a river corridor that connects them to nature and to their heritage. We can gain a great deal of knowledge from people who are maintaining and enhancing connectivity on the ground, so this book includes a variety of such projects to demonstrate important lessons.

In summary, new approaches and policies are needed to curtail land-use change and minimize its impacts. As Albert Einstein wisely noted in another context, "The problems that exist in the world today cannot be solved by the level of thinking that created them." Restoring connectivity is one approach to addressing problems associated with habitat fragmentation. With continued land-use change resulting in habitat loss and fragmentation around the world, we want to encourage people to reconnect their landscapes for the benefit of biodiversity conservation and future generations. Connecting our landscapes can also increase community resilience by providing an opportunity for neighbors to work together on improving their local environment. Through such actions, we may better educate our children about the critical value of conservation for sustaining both natural and human systems.

Chapter 2

Understanding Fragmentation

Habitat loss and consequential fragmentation are considered by many scientists to be the largest threats to preserving the world's biodiversity and a major cause of extinction today (Noss 1987, A. F. Bennett 1999, Henle et al. 2004). Chapter 1 discusses causes of habitat loss, but much of conservation biology is focused on examining the consequences of fragmentation for species' persistence and exploring options such as improved connectivity to enhance survival of native species. In this chapter we examine the effects of habitat fragmentation on biodiversity.

Habitat loss involves reducing the size of habitat and may result in breaking habitat into patches; the latter process is referred to as fragmentation. *Fragmentation* is defined as the transformation of a continuous habitat into habitat patches that vary in size and configuration (Fahrig 2003; Figure 2.1). The effects of habitat loss and fragmentation have not always been carefully parsed out. Cumulative research indicates that habitat loss has consistent negative impacts on biodiversity. Fragmentation, as will be discussed, can have positive or negative impacts, depending on the circumstances and the species of interest (Fahrig 2003). Both habitat loss and fragmentation can be natural or anthropogenic. We focus on human-induced fragmentation in this book but include a short comparison of the differences between natural and human-induced fragmentation.

Temporal and Spatial Scales

Temporal and spatial scales critically affect how we think of habitat fragmentation, as well as corridor conservation efforts, and influence the in-

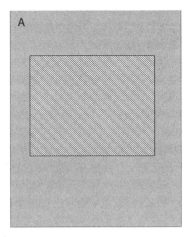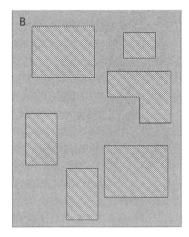

Figure 2.1. Two schematic representations of core habitat (dark) surrounded by matrix (light). The six habitat fragments depicted in B represent the equivalent amount of habitat as in the continuous core in A.

ferences we can make about landscape patterns and processes. Time proceeds on a one-way linear dimension, while spatial scale refers to the two or three dimensions of an object or process and is characterized by both grain and extent. *Grain* is defined as the finest level of spatial resolution possible within a given data set, and *extent* is the size of the study area (Turner et al. 1989; Figure 2.2). Geographers use cartographic scale, in which "large scale" refers to a relatively small study area; whereas ecologists commonly view large scale as equivalent to large geographic areas. In this book, we use *coarse* and *fine* to distinguish differences in grain and *large* and *small* to distinguish extent. While coarse-scale habitat models may predict the presence of species broadly across a region, more fine-scale analyses indicate that species' distributions are often not homogenous but instead are patchy within a region.

Less confusion seems to exist in referring to temporal scale, as most fields reference events as occurring over the long or short term. However, the reference used to define short and long periods is dramatically different among scientific disciplines. Geology addresses processes that occur over millions of years; ecology is often restricted to relatively short-term multiyear phenomena. A significant temporal issue in ecology concerns measuring presence of a species. One-time surveys may indicate that a species is present, but unless the surveys are conducted over multiple seasons and in multiple years, there is no guarantee that the species permanently occupies

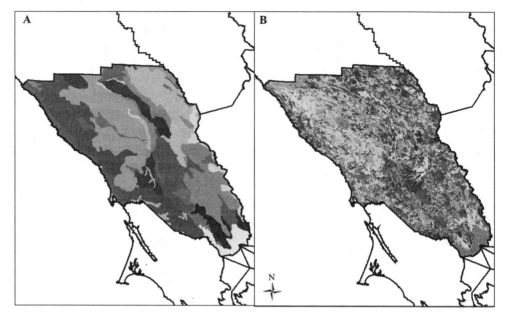

Figure 2.2. Soil types represented by different shades, mapped at a coarse resolution on the left (1:250,000 digital data) and a finer resolution on the right (1:20,000 digital data) for the same geographic extent (4,580 square kilometers [1,786 square miles] in Sonoma County, California).

the surveyed area. Such complicating factors mean that even though a habitat patch or corridor has been established and may initially appear to encompass a focal species, it may not be sufficient for long-term species survival.

Natural versus Human-Induced Fragmentation

One of the greatest limitations in our understanding of fragmentation concerns the differences between naturally heterogeneous and human-fragmented landscapes. Whereas some species adapt to natural fragmentation, it appears that many suffer adverse consequences from human-induced fragmentation. It is because of that difference that maintaining and recreating connectivity has become a central issue on many conservation agendas. Yet we need to understand how we humans adversely affect long-term conservation of biodiversity to mitigate the impacts. Do natural boundaries between intact and burned forests or between avalanche zones and montane forests, as examples, function similarly to those boundaries created by humans? What are the similarities and differences?

There are commonalities between naturally fragmented and human-fragmented systems. Regardless of the mechanism, smaller patches of habitat contain fewer species as well as fewer specialists (i.e., species that depend on specific habitat, foods, or other limiting factors to survive) (Laurance et al. 2002). For example, herb communities that occupy the naturally patchy serpentine soils of California are likely to have a lower diversity of endemic species (species found nowhere else) in small patches than in larger patches (Harrison 1999). Similarly, research in a rain forest fragmented by human activities in New South Wales illustrated that smaller patch size was correlated with lower overall species richness, as well as lower diversity of native species of ground-dwelling small mammals (Denstan and Fox 1996).

In Chapter 3, we discuss the theories concerning the relationships between patch size and species composition in all types of fragmented landscapes. Apart from the similarity just described, human-induced and natural fragmentation often differ in three important ways: (1) speed and pattern of change, (2) scale of change, and (3) ability of the resulting fragments to recover from perturbations.

Speed and Pattern of Change

Natural fragmentation involving large areas tends to occur slowly, in contrast to the rapid human-induced fragmentation. Glaciation, for instance, causes slow fragmentation across large areas. In cycles of about 100,000 years, glacial periods have been separated by approximately 20,000-year interglacial periods (DeChaine and Martin 2004). Evidence suggests that many once continuously distributed populations of species became fragmented into multiple sites or habitat refugia during glaciation. Species in refugia evolved separately, contributing to species differentiation over time (e.g., Brown 1971, Knowles 2001, DeChaine and Martin 2004). During interglacial periods, refugia often become reconnected. On the other hand, the rapid human-induced habitat loss and fragmentation affecting much of the globe has largely occurred in recent history (Sanderson et al. 2002; Figure 1.1).

Many natural processes that cause fragmentation, ranging from fire and avalanches to volcanoes and windthrow, occur with different frequency and result in an altered landscape mosaic than human-induced fragmentation. The impacts of these natural events are often significantly different than those of human-induced changes on the same ecosystem. Tinker and colleagues (2003) compared landscape dynamics in Yellowstone National Park, where natural fires have been allowed to burn, to those in the adjacent Targhee National Forest, which experienced fire suppression along with intense clear-cutting over the past thirty years (Figure 2.3). The researchers found the post-harvest spatial characteristics in Targhee to be very different from the spatial characteristics created by the natural fires (the relatively small pre-1988 fires and the large 1988 fire) in Yellowstone. For example, the pre-logging number and size of forest patches were similar between Yellowstone and Targhee; post-logging areas in Targhee contained significantly more and smaller patches than Yellowstone (Tinker et al. 2003).

Understanding such differences is important in directing future management and restoration efforts, including those of wildlife corridors. In general, the landscape mosaic that results from human-caused fragmentation—from clear-cutting of forests, plowing of grasslands, construction of reservoirs, and development of agricultural and urban artifacts, for example—is different from the mosaic that results from naturally occurring types of fragmentation. The differences in fragment patterns have

Figure 2.3. Aerial photo showing the western edge of Yellowstone National Park (left), which is protected from logging, and Targhee National Forest (right), where the forest has been cleared for timber over the last fifty years. (Photo by Tim Crawford.)

consequences for biodiversity conservation and implications concerning the need for connectivity.

Scale of Change

In addition to the differences in the speed and pattern of change between natural and human-induced fragmentation, humans have contributed to substantially more habitat loss and fragmentation, particularly in the last century, than would naturally occur. That difference can best be seen by comparing the quantities of human-caused fragmentation and natural fragmentation in different biomes. Fragmentation due to human-induced land-use change covers 49 percent of the original tropical and subtropical dry broadleaf forest range, whereas only 6 percent of that same forest type is naturally fragmented. The amount of human versus natural habitat fragmentation is 53 to 4 percent for temperate broadleaf and mixed forest and 55 to 17 percent for Mediterranean systems. With the single exception of boreal forests, where the ratio is 4 to 13 percent, human-induced fragmentation is higher than natural levels of fragmentation in every major biome in the world (Wade et al. 2003).

Ability of Fragments to Recover

Another major difference between human-induced and natural fragmentation concerns the recovery of fragments from disturbance. Areas impacted by naturally caused fragmentation, such as from floods scouring a valley bottom, usually are adapted to go through succession and recover their ecological integrity if left undisturbed. Natural landscape patterns will be reestablished based on soil type and topography following perturbations if left to natural processes.

Human-induced fragmentation is often irreversible, and natural regeneration may not be allowed to occur. Even where human activities such as logging do allow for regrowth, frequent and repeated human disturbances can mean that forests never return to climax or old-growth states. Even selective logging ultimately can alter forest composition and function (e.g., Carey and Johnson 1995, Hall et al. 2003). Often humans alter the original community, including both biotic and abiotic components, so drastically that the original community cannot be restored by natural processes (e.g., Foster et al. 1998, Compton and Boone 2000, Donohue et al. 2000). Sometimes impacts made by humans wipe out propagule sources from which native plant species would reestablish, and topsoil erosion can prevent successful reestablishment even when vegetative propagules do arrive. Further, persistent and regular perturbation can diminish a region's resistance to invasion of exotic species (species that are not native to the region) such that regeneration is dominated by nonnative flora and fauna. Because human-caused fragmentation is often relatively permanent, ecologists and planners have looked toward corridors as a potential tool to retain species in such human-impacted landscapes.

Consequences of Human-Induced Fragmentation

Scientists have been working to understand the impacts of human-induced fragmentation on the world's biodiversity. Cumulative work that has been achieved so far indicates that consequences of human-induced fragmentation for native flora and fauna are extensive. Impacts range from a decline in the numbers of species, population sizes, and species' ranges to increases in exotic species and predation on native flora and fauna (Beier 1993, Stefan 1999). Here we describe some of the factors that influence biodiversity survival in fragmented landscapes, and in Chapter 3 we address theories and mechanistic explanations of why these patterns occur.

The configuration of the landscape after habitat loss determines the impact of resulting fragmentation on the area's original biodiversity. Habitat loss from human activities may result in one remnant patch or multiple habitat patches. The number of patches tends to decline as a result of the original community being progressively destroyed (Figure 2.4). Two of the most important factors influencing species survival in retained patches include the size of the retained patch and the potential or realized connectivity among patches. Other related factors that affect species conservation include edge effect, species loss, and genetic consequences. Good connectivity may ameliorate some of these impacts and allow a species to utilize more than one habitat patch either on an individual level or at the population level, thus helping to maintain regional species' persistence.

Size, Number, and Isolation of Habitat Patches

Larger patches generally harbor more species than smaller patches (Laurance et al. 2002, Steffan-Dewenter 2003, Fahrig 2003), which translates to a higher probability of conserving more native species. The Biological Dynamics of Forest Fragments Project (BDFFP) was initiated in 1979 to assess the effects of fragmentation on biota in Amazonia. Researchers' findings, not surprisingly, often showed a positive correlation between frag-

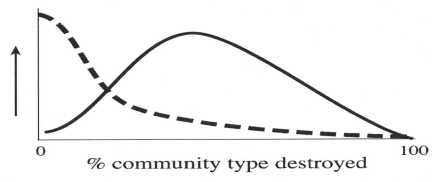

Figure 2.4. Schematic representation of how average fragment size and number of fragments change as the percent of a particular community type in a landscape progresses from complete cover (none destroyed) to 100 percent destroyed. The y-axis is numbers or size. The dashed line is the mean size of fragments, and the solid line is the number of fragments.

ment size and species richness such that intact forests contained more rain forest species per unit area than fragments. Extinction rates were negatively correlated with fragment area, such that more species went extinct within smaller fragments (Laurance et al. 2002). Other research results from BDFFP found that species also were lost from the larger fragments studied (e.g., 100-hectare, or 247-acre, fragments). These extinctions happened over relatively short time spans, indicating a need for immediate conservation action in fragmenting landscapes (Ferraz et al. 2003). Cumulative research from the BDFFP suggests that rain forest species, ranging from mammals and birds to beetles and butterflies, are sensitive to fragment area and that responses to fragmentation are individualistic by species, with some species responding negatively and others responding positively (Laurance et al. 2002).

While retaining larger fragments may help some species survive, retaining only one large residual fragment of a formerly continuous natural community may be risky (Box 2.1). Disease, a catastrophic storm, or some other stochastic event could potentially wipe out all remaining individuals in a single habitat patch. Unfortunately, all the individuals of numerous species currently exist only in a single patch. These include the Gar lions of India discussed in Chapter 1, for example (Oza 1983).

Extensive literature demonstrates that larger fragments that are closer or more connected to source populations have higher species richness than smaller habitat fragments that are more isolated (e.g., Hanski et al. 2000).

Box 2.1.

The SLOSS Debate

The desired number of patches and their size for biodiversity conservation are encompassed in the SLOSS (single large or several small) debate (Rebelo and Siegfried 1992). Given the limited amount of conservation resources, should we be focusing on one large reserve or several smaller reserves in an area (Figure 2.1)? The debate has raged, with proponents of the single large reserve emphasizing survival of the area-demanding species, survival following catastrophic events, and the like. Proponents of multiple reserves contend that each reserve is likely to have a unique set of species and that more species can therefore be conserved by setting aside multiple reserves (Diamond 1976, Simberloff and Abele 1976). They argue further that it is preferable not to tempt extinction by having all individuals of a given species in a single patch, in case an event such as a disease were to eliminate all the individuals in that patch. There is no one answer to the SLOSS debate, as each species will respond differently to the size, number, and location of patches. The debate has simmered down in recent years in recognition that the best plan is to set aside as many reserves as possible and that they should be as large as possible.

In part this is because the relative size and isolation of habitat patches can influence species' persistence, as discussed in more detail in Chapter 3 (Åberg et al. 1995, Dunn 2000; Figure 2.1). Ultimately, many species that require large areas to maintain functional populations will need to move among remaining habitat patches to survive, whether many small patches or several large patches remain. The location of patches relative to one another and the connectivity among patches will play a critical role in their survival. Isolation of habitat fragments from one another can ultimately lead to population declines. Some species may be able to survive as metapopulations in multiple habitat fragments but only if sufficient interchange among the patches occurs (Chapter 3; Fahrig and Merriam 1994). Researchers have documented local extinctions of species in small habitat patches where access to large core habitat areas (habitat not affected by surrounding communities) or other habitat fragments have been cut off. For example, mountain lions (*Puma concolor*) in the Santa Ana Mountains of California have already become extinct in a 75-square-kilometer (29-square-mile) habitat fragment and are expected to become extinct in another 150 square kilometers (58 square miles) of habitat if a housing project severs possible connections with other fragments (Beier 1993; Figure 2.5). Similarly, research in the Northern Territory of Australia showed that rose-crowned fruit doves (*Ptilinopus regina*) and pied imperial pigeons (*Ducula bicolor*) were particularly affected by isolation from other rain forest habitat (Price et al. 1999). Fewer habitat patches within a 50-square-kilometer (20-square-mile) area resulted in fewer species of doves and pigeons within any particular area. This result indicates that the presence of those birds is likely influenced by the presence of other habitat patches nearby. Likewise, red-backed voles (*Clethrionomys californicus*) living in remnant forest patches in the U.S. Pacific Northwest were found to be essentially isolated, in that almost no voles were detected outside of patches, suggesting that dispersal among patches is rare (Mills 1996). Preliminary DNA work suggests that voles remaining in isolated patches have less genetic variation than those in large continuous populations, which may

Figure 2.5. The mountain lion (Puma concolor) is an example of a large, wide-ranging species that exists at low densities and is especially susceptible to human-induced fragmentation. This photograph was taken using a remotely triggered camera in Sonoma County, California, 1999. (Photo by Jodi Hilty.)

make them more susceptible to extinction from inbreeding depression and lack of immigrants to help boost populations in fragments (Mills 1996).

Also, distances among remaining habitat fragments may influence species' use of adjacent human-modified habitat (Downes et al. 1997, Perault and Lomolino 2000; Chapter 5). Species may have different spatial scales of movements such that some species may venture farther from remaining habitat patches than others (e.g., Ricketts et al. 2001). Diversity of nearby habitats and structure or cover remaining in the human-impacted areas also may affect species' presence at a given point in the landscape (Wright and Tanimoto 1998, Drapeau et al. 2000, Hilty and Merenlender 2004, Hilty et al. 2006). For example, wolves in northern Italy were most active in agricultural land adjacent to forested areas (Massolo and Meriggi 1998), and the same was found to be true for carnivores in northern California's oak woodland and vineyard landscape (Hilty and Merenlender 2004). Species unable to use heavily modified landscapes at all may be more prone to extinction (Gascon et al. 1999, Bentley et al. 2000).

Species with higher mobility should theoretically survive better in a fragmented landscape and among more isolated patches of habitat. Such species can move between patches and thereby minimize the negative demographic and genetic consequences associated with small populations (see Chapter 3). However, this is not always the case. Many species possessing the physical capability do not move across human-impacted landscapes, particularly when natural habitat patches are far apart. Those species with limited mobility or other behavioral impediments to moving are often restricted to single patches in fragmented habitats and the associated risks of single-patch occupancy can lead to local extinction (Laurance 1991, Gascon et al. 1999). Species such as grizzly bears (*Ursus arctos*) come into conflict with humans—by eating trash, harassing pets, even attacking people. Such conflict leads to their removal and reduced survival in the human-occupied landscape, and hence their interpatch movements (e.g., Haroldson and Frey 2002; Figure 2.6). Despite their ability to travel long distances, bear populations can become isolated in habitat fragments.

Boundary and Edge Effects

A host of effects associated with edge habitat affect the functioning of ecosystems and biodiversity. Edges are often the location of interactions that influence or limit biotic units (Murcia 1995, Ries et al. 2004; Chapter 5). They are the boundaries between places perceived by an organism

Figure 2.6. An Idaho license plate reflecting anti–grizzly bear sentiment. (Photo by Joel Berger.)

to be significantly different from each other. While edges can separate two or more different communities, they are not necessarily obvious to the human eye and can be difficult to measure in width or length (Lidicker 1999).

Although edges occur naturally, this discussion of edge effects will be limited to the consequences of human-created edges (e.g., see Figure 2.3). The division of one continuous natural habitat by humans into one or more smaller fragments of habitat results in a human-created edge where the natural habitat ends and abuts the human-altered parts of the landscape. Often, this is a hard edge, at least initially, with little or no successional or transitional stage such as one may see in the natural environment (see Figure 4.2). Hard-edged boundaries have a stronger negative impact compared to more natural transitional edges (Mesquita et al. 1999). In addition, the shape of the retained habitat affects the ratio of edge to core habitat, such that perfectly round habitat fragments have the smallest edge-to-area ratios; more irregularly shaped patches have more edge per unit area (Figure 5.3; Lidicker 1999, Laurance et al. 2002).

Why does the amount of edge for a given habitat patch alter ecosystem function and structure within a patch or a corridor? Both physical and biological changes are associated with edges (Ahern 1995, Puth and Wilson 2001, Laurance et al. 2002, Ries et al. 2004). Microclimate alteration is a physical effect of edge that can negatively affect forest interior species while positively affecting other species. At the edge of a forest, direct sun increases, wind exposure is higher, and snow loads can increase (Laurance 1991, 1997, Forman 1995). In forested fragments, soil temperature also increases to resemble that in nonforested areas (Mills 1996), and overall temperature and vapor pressure can vary more along edges. Humidity and depth of humus also often differs on edges compared to interior habitat (Mills 1996, Carvalho and Vasconcelos 1999). Such influences can extend as much as 50 meters (164 feet) into forest patches (Laurance 1997, Sizer and Tanner 1999). These altered conditions can inhibit regeneration of vegetation where seeds are particularly sensitive to desiccation and can increase mortality due to windthrow (Laurance 1991, 1997). As a result of these microclimatic alterations, vegetation composition often is different at the edge compared with the interior of a forest, and altered vegetation can be detected up to 500 meters (1,640 feet) into a tropical rain forest (Laurance 1995). In Amazonia, tree species along the edge of habitat fragments suffered high mortality (Mesquita et al. 1999). For very small fragments of natural habitat or narrow corridors, microclimatic changes associated with the edge may permeate throughout.

Such changes in microhabitat and consequently to natural vegetation can be one of the contributing reasons for corresponding faunal changes in composition and density. Strongly edge-intolerant species may have higher densities in interiors compared to edge areas (e.g., Mills 1996). Such negative edge effects may ultimately result in smaller functional habitats within retained remnants. If a patch is too small, the entire patch could be considered edge habitat, thus eliminating favored habitat altogether. Likewise, narrow corridors may not serve such species.

In contrast, some species benefit from or prefer edge and increase in abundance there. Edge offers access to resources within multiple habitats, helping a number of different species to thrive (Ries et al. 2004). The altered abundance or distribution of species near the edge can further affect species within habitat patches through negative interactions (Murcia 1995). Generalist predators and exotic species are often edge species and sometimes out compete specialists and native species. They can also contribute to increased predation, competition, and parasitism on native interior species (Beier 1993, Murcia 1995, Stefan 1999). Smaller habitat fragments with higher edge-to-area ratios provide increased access of weedy species into fragments and can enhance movement of edge-loving exotic species and pests (Panetta and Hopkins 1991). Brown-headed cowbirds (*Molothrus ater*), raccoons (*Procyon lotor*), opossums (*Didelphis virginiana*), and crows (*Corvus* spp.) are examples of species that thrive in edge habitat and can have a large impact on forest interior species. Such species act as nest predators, nest parasites, or cavity competitors of interior species, and they can contribute to decreased populations of ground-nesting birds, forest songbirds, reptiles, and amphibians in remaining habitat fragments (Paton 1994, Harris et al. 1996, Hartley and Hunter 1998, Dijak and Thompson 2000, Hansen et al. 2002). In the Purcell Mountains of British Columbia, increasing fragmentation and logging contributed to edge-loving deer (*Odocoileus virginianus*) moving into a region not previously occupied by deer, followed by mountain lions (Figure 2.5). The mountain lions now have found mountain caribou (*Rangifer tarandus caribou*), a naive prey and easy target, and are significantly contributing to the declining numbers of an already tenuous caribou population (Kinley and Apps 2001).

Species associated with humans can also reduce native biodiversity in retained fragments by invading the edges and smaller fragments (Stefan 1999). Domestic and feral animals, such as cats and dogs, damage native species populations in remaining habitat by chasing and preying upon them (Simberloff and Cox 1987, Ahern 1995, Laurance 1995, Arango-Velez and Kattan 1997, Crooks and Soulé 1999). Similarly, pioneer plant species that did not previously occur cause problems for the original community of species. For example, such plants invaded up to 10 meters (33 feet) into tropical rain forest patches in Brazil (Sizer and Tanner 1999). Likewise, livestock can encroach upon remnant habitat and either directly compete with other herbivores or alter the habitat characteristics, leading to decreased use by wildlife (D. A. Norton et al. 1995, Kemper et al. 1999). Fragments can also experience direct increased human impacts, especially when adjacent to high human density. Increased edge allows humans greater access for recreation, including legal and illegal hunting of animals (Simberloff and Cox 1987). The habitat fragments may then become devoid of species that are sensitive to human activity or heavily hunted, which can have cascading impacts on the remaining ecological community.

Adjacent human activities can also affect the overall integrity of remaining habitat patches (Perault and Lomolino 2000). In agricultural zones, pesticides and fertilizers can drift from the fields into habitat and affect flora and fauna alike (Forman 1995). In addition, research on small fragments in South African renosterveld shrublands suggests that grazing, trampling, and fires in the human-occupied lands affect the remnant habitat (Kemper et al. 1999). Adjacent roads also can pollute retained habitat. For example, nitrogen deposition from air pollution threatens native grasses and associated biodiversity on serpentine soils in northern California. Because nitrogen is a limiting nutrient in serpentine soils, deposition of nitrogen changes the composition of grasses, facilitating generalists and contributing to a decline of serpentine-associated grasses and forbs (Weiss 1999).

In addition to fragmenting habitat, roads can also be a source of light, noise, and mortality for mobile species that occupy adjacent habitat patches. Studies indicate that light pollution can disorient animals, such as turtles, often imperiling species (e.g., Tuxbury and Salmon 2005). The combination of light, noise, and high human activity can deter the presence of some species in adjacent habitats. For example, female grizzly bears in Alberta, Canada, showed a negative relationship to areas with more vehicles, traffic noise, and human settlements. Male grizzly bears in the same study were more likely to use high-quality habitat near roads at night, especially where cover existed (Gibeau et al. 2002). While many species show a tendency to avoid roads with high vehicle usage, roadkill remains a large source of mortality for wildlife. Vehicle collisions with large mammals are increasing in developed countries, and several million collisions per year occur worldwide (Malo et al., 2004). Researchers in Saguaro National Park on the U.S.-Mexico border estimate that fifty thousand animals, including reptiles, amphibians, birds, and mammals, are killed by vehicles annually (Araiza 2005). Dispersing and young individuals of many species may be particularly susceptible to becoming roadkill due to their inexperience, a factor that may be important to consider for rare or endangered mobile species. For example, automobiles killed four of nine radio-collared mountain lions observed dispersing in Southern California (Beier 1995). Species occupying patches next to or bisected by roads also have an elevated risk of mortality. Florida scrub jays (*Aphelocoma coerulescens*) with home ranges adjacent to roads had a significantly higher mortality rate than jays far from roads (Mumme et al. 2000).

Edge effects can contribute to changes in original communities. Small habitat patches and narrow corridors are more susceptible to such impacts, and that should be factored into corridor designs.

Species Loss and Change in Composition

Edge effects along with other impacts can ultimately contribute to loss or turnover of species in remaining habitat patches, which in turn may alter the performance of the system (A. F. Bennett 1999, Rosenblatt et al. 1999, Nupp and Swihart 2000, Naeem et al. 1994; Figure 2.7). Weedy species and generalist predators become more prevalent, while rare and sensitive

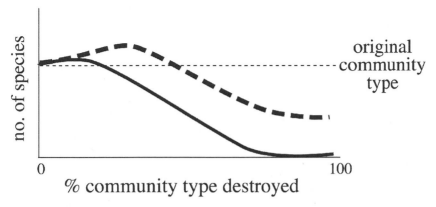

Figure 2.7. The fate of species diversity in a landscape as a once continuously distributed type of community is progressively destroyed. At first, there is an increase in the number of species as new species move into the landscape; some species may actually invade the original community of patches; others will live in the newly created habitats. Eventually, the original community disappears and only the species associated with the new habitats persist. The solid line represents the number of species in the original community type; the dashed line is the sum of the original species plus new ones that enter as the landscape changes.

species tend to decline (Brosset et al. 1996, Dijak and Thompson 2000). Research in the Atlantic forest of southeastern Brazil suggests that there is a shift in plant structure and species composition as tropical forest fragments are reduced in size, with small fragments containing more weedy plant species (Tabarelli et al. 1999). Similarly, a study in the South African renosterveld shrublands shows that the number of exotic plant species increases with decreasing size of habitat fragments (Kemper et al. 1999).

While weedy species may appear or increase in abundance in habitat fragments, species that were once widespread may disappear. The loss of species may then lead to multiple additional impacts. A good example of this was documented in western Australian eucalyptus (*Eucalyptus salmonophloia*) woodlands where mistletoes (*Amyema miquelii*) were more likely to be found in large fragments of habitat than in small fragments (D. A. Norton et al. 1995). This is likely due to the sensitivity of both mistletoe and avian dispersers to fragmentation. Loss of mistletoe in fragments ultimately may contribute to a decline in two butterfly species (*Ogyris* sp. Lycaenidae and *Delis aganippe* Pieridae) that feed on mistletoe, thereby contributing to a further cascade of species loss in these habitat fragments (D. A. Norton et al. 1995).

Like plants, native fauna often disappear from small patches, which can also lead to cascading effects (Box 2.2). Large predators are often the first species to go extinct as fragment size diminishes (Beier 1993, Mac Nally and Bennett 1997, Primm and Clark 1996; Figure 2.5). Terborgh and his colleagues (2001) showed that the loss of large predators began an ecological meltdown on islands created by dams in Venezuela. Consequent to the loss of predators, the researchers documented that herbivores increased greatly, leading to a cascading loss of much plant and other animal diversity. Parallel to the Venezuelan study, the unraveling of an ecosystem was documented on the Pacific Coast of North America, where sea otters (*Enhydra lutris*) disappeared, then herbivores became overpopulated, and kelp forests began to disappear (Estes and Duggins 1995). In other systems, such as in Southern California, loss of large carnivores has led to the release of pressure on middle-sized predators, with consequent shifts in the dynamics at herbivore and plant trophic levels (Crooks and Soulé 1999).

After isolated islands or fragments of habitat are created, species loss can occur gradually over time, which is sometimes called *species relaxation*. A good example of this phenomenon comes from lizard fauna that experienced cumulative loss in species diversity over time on islands in Baja California that were once linked to the mainland and are now isolated (Wilcox 1978). Even national parks may lose species as parks become more isolated due to surrounding human activities. Newmark (1995) documented that species loss was greater than species recruitment in western North American national parks. Recent research shows that human-caused habitat fragments also experience species relaxation. Older chaparral and coastal sage scrub habitat fragments in urban San Diego County, California, contained fewer chaparral and scrub specialist bird species (Bolger et al. 1997, Crooks et al. 2001). Likewise, bird species occurring at lower densities were more likely to go extinct than species with higher densities. Studies documenting species relaxation are important, because they indicate that species currently found in habitat fragments, especially relatively new fragments, may not be able to survive in those fragments in the long term.

Small isolated populations can experience genetic erosion, inbreeding, and reduced fitness that can ultimately lead toward population extinctions. A long-term study of an isolated population of greater prairie chickens (*Tympanuchus cupido pinnatus*) in Illinois documented both decreased fertility and decreased egg hatching rate, with a decline in genetic diversity (Westemeier et al. 1998). When translocated birds from other populations

BOX 2.2.

A Sample of the Impacts of Habitat Fragmentation on Birds

Because birds are comparatively easy to study, much research effort has been directed toward examining different impacts of habitat fragmentation on birds. Impacts of fragmentation on birds, as with most species, are often confounded by concomitant habitat loss. Few studies have successfully teased apart the effects of fragmentation and reductions in habitat availability (Schmiegelow and Mönkkönen 2002, Fahrig 2003). Despite that caveat, cumulative research indicates that there are a variety of impacts but that they are often community or species specific. Some examples of changes in bird communities include the following:

- Less overall available habitat generally means lower species richness and fewer specialists (e.g., Drapeau et al. 2000, Schmiegelow and Mönkkönen 2002).
- Larger habitat patch sizes will have more individuals as well as species of birds, with some species being absent from smaller patches altogether (e.g., Beier et al. 2002). Some reasons for absence from smaller patches may be food shortages and limited nest sites (Burke and Nol 1998, Zanette et al. 2000).
- Habitat fragmentation can impact bird density and fecundity, which may be influenced by patch size, distribution of patches across the landscape, and landscape composition (Donovan and Lamberson 2001, Donovan and Flather 2002).
- Bird species' distribution in fragmented habitat is species specific; some birds occur across all patch sizes and throughout patches, while other birds may be more likely to appear only in certain patches sizes or only in the interior or the edge of patches (e.g., Trzcinski et al. 1999, Ford et al. 2001, Beier et al. 2002, Schmiegelow and Mönkkönen 2002).
- Edge effects can affect avian nesting success (Paton 1994, Batáry and Báldi 2004). The impact of edge effects, including increased nest parasitism and nest predation, is community, species, and scale specific (e.g., Bolger et al. 1997, Beier et al. 2002, Chalfoun et al. 2002, Stephens et al. 2003). The type of vegetation and the location in which birds nest can affect predation rates (e.g., Aquilani and Brewer 2004, Borgmann and Rodewald 2004). Overall landscape composition also influences nest predation (Hartley and Hunter 1998, Zanette and Jenkins 2000).

were introduced into the isolated study group, increased fertility and egg hatching rates resulted, indicating that low genetic diversity is impairing population recruitment. Similarly, song sparrow (*Melospiza melodia*) reproduction was studied on islands of various size and isolation from the North American continent, where researchers documented that natural selection favored noninbred birds after the populations experienced bottlenecks or population crashes (Keller et al. 1994). These studies are examples illustrating that low genetic diversity can be a major factor contributing to a decline in species. Because of such evidence, researchers are showing that conservation efforts may be most successful when focusing on larger populations. For example, research indicated genetic erosion in small populations of an endangered tetraploid pea (*Swainsona recta*) in Australia, such that conservation efforts should focus primarily on populations of fifty or more reproducing plants rather than the smaller inbred populations (Buza et al. 2000). One way to diminish genetic inbreeding impacts and boost effective population size is through reconnecting disjunct populations, as through corridors.

While genetic issues contribute to population declines, genetic erosion is not always the ultimate factor causing extinction for many mammal species. Extirpation can also be the result of a change in the environment that may reduce a population size or distribution. That may lead to extinction due to demographic collapse, as when a small population lacks the requisite breeding-age adults of each sex, or environmental stochasticity, such as a large storm, fire, or disease (Gilpin and Soulé 1986). Small populations may lack a reproductive-age female or male, or a drought or some other natural environmental variation may eliminate remaining straggler populations. For example, only small populations of black-footed ferrets (*Mustela nigripes*) remained in the world by the 1980s. An outbreak of canine distemper nearly wiped out the species, and ultimately the population plummeted to eighteen in captivity, probably all that remained of the species (Seal et al. 1989). In that case, the population escaped total extinction because of human intervention, but it illustrates how small populations are vulnerable to disease and other natural catastrophes. That and other ecological processes that can cause problems for species' persistence are discussed in more detail in the next chapter.

In summary, human-induced fragmentation can happen quickly across large areas and can permanently affect biodiversity. Fewer, smaller, and more isolated habitat patches with increased edge effects can lead to species loss and changes in community composition.

CHAPTER 3

The Ecological Framework

Organisms on planet Earth are not evenly distributed across its surface. The tremendous variety of conditions found from the deepest oceans to the tops of the highest mountains, from the poles to the equator, and around various latitudes ensures that the various kinds of organisms will be discontinuously distributed. On smaller than global spatial scales, species are generally limited to certain continents or biotic provinces within continents and to places that suit their adaptations. It is sometimes the case, moreover, that what constitutes suitable habitat will vary with stage of development or season. A frog, for example, may need a pond while it is a tadpole and a forest as an adult. A sandpiper will use Arctic tundra in the summer for breeding and a tropical mudflat in the winter.

Organisms have always had to deal with discontinuous habitats, as well as their own changing needs over both short and long terms. What is new is the accelerating rate of habitat fragmentation that has been occurring over the last few hundred years in response to human population growth and technology advances. Increasingly, organisms are confronted with disappearing and severely fragmented habitat, as well as exotic predators, diseases, and competitors, making it difficult for them to cope effectively with inevitable catastrophes and rapidly changing world climates. In this chapter, we review the ecological concepts that address these challenging issues of spatial discontinuity to which organisms (individuals, populations, and communities) are being forced to adapt. (See *Spatial Ecology, the Role of Space in Population Dynamics and Interspecific Interactions*, edited by Tilman and Kareiva, 1997, for more detailed treatment of many related topics.)

What Is Connectivity?

Connectivity is a measure of the ability of organisms to move among separated patches of suitable habitat; it can be viewed at various spatial scales (see P. D. Taylor et al. 1993). Continents can be connected and disconnected with varying sea levels. North America and Asia, for example, were periodically connected by a Siberian-Alaskan land bridge during the Pleistocene. In this book, however, we will concentrate on connectivity between patches of suitable habitat that are potentially relevant to the experiences and welfare of individual organisms. Since organisms vary tremendously in their abilities to travel (vagility) and in their motivation to leave their home area, the degree of connectivity in a given mosaic of biotic communities will vary greatly according to the perspective of each species.

This brings us to a consideration of what is meant by corridor. As you will discover throughout this book (especially in Chapter 4), the term *corridor* is used in a variety of ways. For this book, we define it as any space, usually linear in shape, that improves the ability of organisms to move among patches of their habitat. It is therefore important to recognize that what serves as a corridor for one species may not be a corridor or may even be a barrier to another. Corridors can be natural features of a landscape or can be created by humans.

Island Biogeography

The first major effort to theorize about how organisms deal with disconnected patches of habitat was the theory of island biogeography. This perspective was formalized in a now classic book by R. A. MacArthur and E. O. Wilson, *The Theory of Island Biogeography*, published in 1967, which triggered a flood of interest among ecologists in this topic and energized much research effort. The theory in its original form dealt with real islands, bits of land surrounded by water. The focus was not on how individual organisms might move among the islands and the mainland, but on what influenced the diversity (or number) of species on islands. The particular species involved were not of direct concern; rather, the number of species or species richness was.

To address this interesting and important question, four simple propositions were made:

1. Larger islands will host more species than smaller ones because large islands are likely to have more different habitats on them. Greater topographic diversity will lead to more microclimates and to more soil types, so that a greater diversity of plants and micro-organisms can live there, and will thus support a larger variety of animals, as well. Furthermore, large islands will present larger targets for potential colonizers to find.
2. Islands close to the mainland will be more diverse than more distant islands. The mainland is presumed to be the source of colonizers to the islands, and so the closer the island is to the source of immigrants, the more likely it is to be reached by them.
3. Small islands will suffer higher rates of species extinctions than large islands. This is because small islands will support generally smaller numbers of individuals for each species present, and residents will therefore be continuously at a greater risk of having their numbers decline to zero. Moreover, species that compete with each other (have negative impacts on each other) will be less likely on small islands to find some microhabitat or spatial refuge where they can avoid each other.
4. Islands close to the mainland will experience lower extinction rates regardless of size because they will benefit from a larger input of new colonists, including individuals of species already present on the island. Thus, even species persisting precariously on a near island will frequently have their numbers enhanced by new arrivals from the mainland. Species on distal islands will only rarely benefit by the arrival of new recruits to their ranks.

The interactions among these four basic principles can be shown in two graphs (Figure 3.1). The resulting species richness on any given island can be seen as an equilibrium between the rate of new species becoming established on an island and the rate of colonizers going extinct. Species number (richness) is therefore a dynamic concept and one that is sensitive to even small changes in colonization and extinction rates. Species richness equilibria will be greatest on large islands close to mainland sources, whereas small or distant islands will have the lowest equilibria. Although the principles involved here are simple and intuitive, just how island distance and size play off each other depends on the details of any given situation. One application of this theory that has been generally neglected is to aquatic "islands" in a "sea" of land. By analogy, large lakes close to

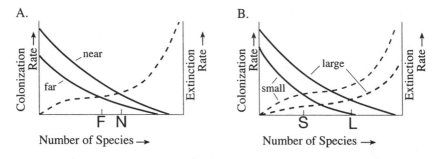

Figure 3.1. Graphic depiction of species richness on islands as determined by distance from the island and island size. Solid lines represent the rate of colonization by new species; dashed lines are species extinction rates. Except on large islands, extinction rates increase rapidly at very low species richness because communities will be simple at that stage and may lack elements that new colonists need. Graph A shows the number of species on islands of a given size as a function of distance from source populations. F and N are equilibrium values for far and near islands, respectively. Graph B shows the number of species on islands at a given distance from mainland sources as a function of island size. S and L are equilibrium values for small and large islands, respectively.

other lakes should have richer biotas than small lakes well isolated from others.

Because of its primary focus on species richness, the theory of island biogeography ignores a number of interesting and important issues related to colonization and survival on islands—for example, the fact that island biota will be biased toward organisms with particular morphological or life history traits. Specifically, island colonists are likely to be species with good vagility—that is, those that can fly, swim long distances, or float easily in water or air currents. Colonization will also be favored by species in which the smallest number of individuals can start a new population. This might be a single seed, spore, or pregnant female. If at least one male and one female are necessary for establishment, the chances of that happening would be the probability of a male arriving multiplied by that of a female arriving at about the same time, a much smaller number than for arrival of a single colonist. Of course, colonization by one or just a few individuals may not provide sufficient genetic variability so that ensuing offspring can successfully adapt to their new environment. If genetic variability is important for success, a large pool of immigrants would be needed. Least likely to succeed would be species that require large social

groups to be successful. For them, enough individuals would have to arrive together, so that their advantageous social behavior could be established. Evidence from the West Indies shows that island life may perpetrate
a physiological bias as well. Wilcox (1978) reported that organisms with
high metabolic rates, such as terrestrial mammals, suffer higher rates of
extinction than do reptiles and amphibians. Flight in birds and bats over-
comes this metabolic disadvantage to some extent. Similarly, large species
characteristically have higher extinction rates on islands than small species.
This is simply because a given island is effectively smaller for large species
than for small ones, large body size generally being correlated with large
home range size.

Other shortcomings of island biogeographic theory are that it does not
consider evolutionary changes subsequent to colonization and does not
allow for speciation either on an island or within an archipelago of islands.
The initial colonists, given sufficient time, can evolve into many species.
The Hawaiian and Galápagos islands are great examples of speciation happening in spectacular fashion. A more subtle evolutionary change is that
the colonist will evolve away from its mainland parental stock (differentiation). This could be by genetic drift (random genetic changes in the
population), by selection acting to improve a species' adaptation to the
new island environment, or both. For example, island populations often
evolve in the direction of losing their dispersal adaptations, such as wings.
After all, individuals that disperse off the island are unlikely to produce
offspring. If island forms lack predators, they will lose adaptations for
predator defense. The fact that large-bodied species suffer a disadvantage
on islands, as mentioned, sometimes translates into evolution of smaller
size, such as occurred with elephants stranded on Mediterranean islands
during the Pleistocene. Moreover, small species, such as rodents, tend to
become larger on islands (Adler and Levins 1994). These evolutionary
responses to island environments, especially well established for mammals,
have been called the island effect (Van Valen 1973) and, as we will see later,
may also apply to terrestrial habitat fragments. If such evolutionary
changes do occur, newly arriving colonists from the mainland may not
recognize the previous colonists as the same species, or they may be re-
jected by them as unsuitable mates. If mating does occur, the resulting offspring may suffer hybrid inviability. In any case, the newcomers' genes
may impede the island population's path toward improved local adaptation; hence those residents who avoid mating with the newcomers will
likely produce more successful progeny.

Another limitation to the theory is that it ignores species-specific demographic behaviors and especially does not consider that these behaviors may have to adjust to a different community composition on the island than in the ancestral home on the mainland. Not only may the mix of competitors a species encounters be different, but there will likely be different predators and parasites to challenge it, as well. Conversely, the absence of familiar predators or parasites may offer different challenges and opportunities. A colonist that finds itself without its usual enemies may face instead the unfamiliar challenges of high-density stresses and food shortages. Finally, it should be mentioned that the focus of island biogeography on island size and isolation ignores the reality that particular air and water currents may well enhance or diminish the probability of potential colonists reaching a particular island relative to others.

Further problems were encountered when attempts were made to generalize island biogeographic theory to terrestrial arrays of fragmented habitats. The theory was applied to so-called sky islands in the Great Basin of North America (Brown 1971, 1978, Brown and Lomolino 1998); these were isolated patches of boreal or subalpine habitats on the tops of the higher mountains in the region. Desert and various semiarid conditions effectively isolated these cooler and moister fragments. Except for birds and other strong-flying creatures, this isolation was generally found to be complete. That is, the colonization rate was zero or close to it, at least for the groups of organisms analyzed in detail, such as terrestrial mammals. These sky island biotas were not in equilibrium between colonization and extinction rates, as prescribed by the theory. They were, in fact, remnants of more widespread species distributions that occurred during cooler and moister periods in the Pleistocene. What we have left is remnant or refugial populations influenced primarily by factors affecting extinction rates, not colonization rates which were effectively zero. Bird diversity on these sky islands fits the classic island biogeographic model better because for them colonization rates are usually above zero.

A general complication of island theory is that strictly speaking it applies only to oceanic islands, that is, to those that emerge from the ocean devoid of terrestrial life. Then the processes of colonization and extinction can proceed as expected by theory. Many islands, however, are land-bridge islands. This means that they were once part of a mainland but became islands through earth movements or rising sea levels. Such islands start life with a sample of the mainland biota already in place. Because these islands tend to be fairly close to a mainland, they will experience colonization events, and therefore a portion of their fauna and flora will behave as prescribed by classical island biogeographic theory. However, these land-bridge islands will also support organisms that cannot disperse over water and so will not be supplemented by colonization events. The continuing presence of such species on the island will be determined only by the probability of extinction. Extinction rates will of course be influenced by island size, topographic diversity, and the chance occurrence of other species by colonization. The richness of this noncolonizing element on land-bridge islands will therefore gradually diminish over time as inevitable extinctions take their course. This process of community relaxation is illustrated in Figure 3.2 by species diversity of lizards on land-bridge islands off Baja California as a function of elapsed time since island isolation (Gonzalez 2000). Island theory projects an equilibrium species richness for islands, but these extinction decay functions, caused by a lack of colonization, introduce a nonequilibrium element into the situation. Understanding the rate at which species will be lost due to island isolation is important because, as we will discuss later, similar phenomena can occur in terrestrial community fragments, as, for example, those produced by forest cutting.

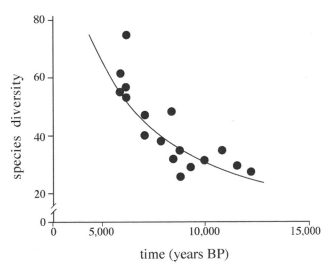

Figure 3.2. Lizard species data from land-bridge islands off of Baja California, Mexico, showing a decline of species richness in the absence of colonization over time since island isolation (data from Wilcox 1978).

The general problem with applying the classic theory to terrestrial habitat mosaics seems to be that they are much more complex than the relatively simple island-mainland models envisioned by this body of theory. In particular, matrix (i.e., various communities located between patches; see Chapter 5), instead of being inhospitable water, can vary greatly in the extent to which it inhibits dispersal. It may, on the one hand, be very inhospitable or even a complete barrier to movements, or it can be temporary habitat or even lower-quality permanent habitat. Matrix properties can also vary seasonally or from year to year. Confronted with these newly appreciated realities, ecologists turned to new conceptual frameworks such as metapopulation theory. Island biogeographic ideas, nevertheless, were critically important in influencing the thinking of ecologists and conservation biologists, and they continue to contribute to our understanding. Rather than being discarded, it is more realistic to think that island biogeographic theory has simply been incorporated into more complex and more generally applicable formulations.

Metapopulation Theory: Conceptual History

The development of the metapopulation concept was a major advance in both spatial ecology and conservation biology. A *metapopulation* is simply a population of populations, or a system of local populations (demes) connected by movements of individuals (dispersal) among the population units. The concept thus recognizes that species are usually arranged in variously disconnected patches across the species' distribution. This fact has been recognized for a long time by naturalists and field ecologists. It was an awkward fact, however, as many ecologists and especially modelers thought of species populations as effectively infinite in size and uniformly distributed throughout the species' range. Obvious exceptions could be conveniently attributed to "unnatural" human influences and dismissed as scientifically unimportant. As habitat fragmentation accelerated in recent decades, the existence and importance of metapopulation structuring became impossible to ignore.

Even before World War II, Soviet ecologists concerned with rodent pest control were emphasizing the disjunct nature of species distributions and the relevance that that had for demographic behavior (Lidicker 1985, 1994). Then in 1970, P. K. Anderson published a paper calling those insights to the attention of ecologists everywhere and applying them to his own research on house mice (*Mus musculus*). In 1969 and 1970, Richard

Levins published two critically important papers in which he modeled metapopulations mathematically. In the second paper, he introduced the term "metapopulation." Interestingly, as with the earlier Soviet biologists interested in rodent control, Levins was motivated by the epidemiological problem of how to control parasites and diseases that were living in a mosaic of habitat patches, namely the bodies of the various individuals that constituted their host species. His insight was to treat each population of pathogens living in an individual host as analogous to an individual in a population. Thus his populations (demes) had birth and death functions just as individuals did. Because of this insight, his mathematical treatment, and the new label (metapopulation), and especially because the time was propitious, Levins's papers represented a defining moment in ecology.

It is important to understand the basics of Levins's model because it requires a number of important assumptions that need to be evaluated, and because it was used as a starting point for development of many more complex and realistic models (Dobson 2003, Hanski and Gilpin 1997). Levins's fundamental equation

$$dp/dt = mp\,(1 - p) - ep$$

expresses how the proportion of habitat patches or hosts (p) occupied by the species in question changes over time (dp/dt); this proportion can vary from zero (extinction) to full occupancy ($p = 1$). In this equation, m is the dispersal rate (migration) of individuals from an occupied patch to one that is unoccupied; mp is therefore the total amount of successful dispersal from all occupied patches and is like a reproductive or birth rate. So the first element in the growth equation expresses the rate of establishment of new colonies in available empty patches $(1 - p)$. The second element is the rate (probability) of extinction of a single occupied patch (e) multiplied by the proportion of occupied patches. This is the death rate (mortality) of the array of occupied patches. Obviously, if the birth of new patches exceeds the death (extinction) of occupied patches, the proportion of occupied patches (p) will increase. If births and deaths are equal, p is a constant, and if the death rate exceeds the birth rate, p will decline toward zero. So, just as with a single population, it is the ratio of losses to additions that determines the growth trajectory of the metapopulation.

A number of critical assumptions should now be apparent for this extremely simple model:

1. Extinction rate (e) is a constant; that is, all demes have the same probability of going extinct, and that rate does not change over time.
2. Dispersal rate (m) is a constant; that is, all demes produce the same rate of successful dispersers to unoccupied patches of habitat.
3. All unoccupied patches ($1 - p$) are equally accessible to dispersers; that is, their actual spatial arrangement is irrelevant.
4. The matrix between the habitat patches is uniform everywhere and its nature is specified only to the extent that it influences m; that is, if the matrix were completely inhospitable to dispersers, m would be zero even if occupied patches were sending out a steady stream of emigrants.
5. All habitat patches are equivalent; that is, they are all of the same size, they can all support the same size population, and their explicit spatial position in the metapopulation array is irrelevant.
6. When a disperser arrives at an unoccupied patch, the patch immediately becomes fully inhabited and begins to produce emigrants; that is, there is no local population dynamics, and patches exist in only two states: + or − (i.e., occupied or unoccupied).
7. There is no dispersal among occupied patches.

Note that metapopulations of this type always have some empty patches, unless $e = 0$. Patches will "blink on and off" as demes go extinct and the habitat patch subsequently becomes recolonized. If m is not less than e, the long-term fate of the metapopulation will depend on the probability of all occupied patches becoming extinct at the same time (e^n), where n is the number of occupied patches. Therefore, continued existence of metapopulations not only will be sensitive to changing vales of the parameters m and e, but also will be dependent on the number of patches, particularly the number of those occupied. Because all of the assumptions implicit in Levins' model would seem rarely to be true, this type of metapopulation is probably rare in the real world. Nevertheless, the model represents a useful abstraction of spatially structured populations. An example of the oversimplification inherent in this minimalist model is given by Hastings (2003), who shows that if the probability of demic extinction is not a constant but increases with patch age (because of succession or impacts of the focal species itself), colonization rates needed to sustain the metapopulation may be twice as much as those determined by the Levins' approach (see also Dobson 2003). Therefore, a dispersal rate that can sustain a metapopulation composed only of young patches will become inadequate as the patches age or become variable in age structure.

Because of the abstract nature of Levins's metapopulation model, it is important that metapopulations not be defined in terms of that model. Rather, a general definition is needed, as we provided previously (a system of variously connected demes of a single species), to accommodate the variety of spatially structured populations of organisms that actually exists. It would be a serious mistake if we restricted our investigation of spatial structuring to a limited subset of what organisms do. Our understanding would inevitably be confined in such a situation to the defined subset of patterns. We would then have to come up with new definitions for spatial patterns excluded from our definition, or remain ignorant of those aspects of the living world. Our approach here is to start with a general definition and then determine if useful subpatterns can be defined. It was in this spirit that Harrison (1991) proposed recognizing four categories of metapopulation (Figure 3.3):

1. *Patchy.* A demic structure is present, but the demes are well connected, and dispersal is common among the populations. If and when a deme becomes extinct, the patch it occupied is quickly recolonized (rescue effect). This is a common pattern and is the one most resistant to metapopulation extinction.
2. *Core-satellite.* This pattern is also called "mainland-island" or "source-sink." It consists of one or more large, extinction-resistant populations (large habitat patch) plus one or more separated, usually peripheral, smaller patches of habitat. Dispersal is mostly in the direction of mainland to satellite. The smaller populations experience occasional extinctions, but eventually those patches are recolonized from the core population. This kind of metapopulation is also relatively extinction resistant because of the relative security of the core or source populations.
3. *Classic Levins.* This is the mathematical modeler's favorite pattern, and its features have already been given. The long-term prospects for such a metapopulation depend critically on the number of patches and the level of connectivity among them.
4. *Nonequilibrium.* This arrangement has the highest risk of overall extinction, because connectivity among the demes is weak or absent, and there is no large secure deme among them. With minimal dispersal among populations, the rescue effect is unreliable or nonexist-

A Patchy

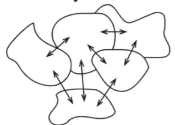

B Core-Satellite

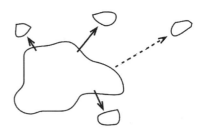

C Levins's Classic

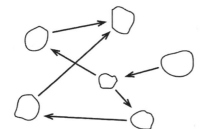

D Nonequilibrium

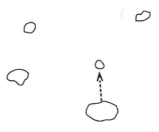

Figure 3.3. Four types of metapopulations as defined by Harrison (1991). Dispersal movements among habitat patches are shown by arrows, and extremely rare dispersal events are depicted as dashed arrows. In Levins's classic model, movements are only from occupied to unoccupied patches, and the spatial arrangement of the patches is not specified.

ent. As demes suffer extinction, there is little prospect of these habitat patches being recolonized, at least over reasonable time spans. Nonequilibrium metapopulations are in a sense moribund. Unfortunately, this pattern is also common and becoming more so with increased habitat fragmentation. Such situations are the ones to which most conservation efforts are directed.

Of course, many species will be composed of an array of metapopulations consisting of a mixture of types. This reality is nicely illustrated by the Florida scrub jay (*Aphelocoma coerulescens*), an endangered species confined to oak scrub habitats on sandy soils in the central part of peninsular Florida (Stith et al. 1996). As of 1996, the entire species was distributed into forty-two metapopulations. The boundaries of these population arrays could be objectively defined because the dispersal distances for the birds are accurately known. The frequency distribution of the number of adult pairs of jays in these metapopulations is shown in Figure 3.4. Note that half of the metapopulations consisted of ten or fewer pairs. Only four metapopulations contained the relatively safe number of three hundred or more pairs. Assigning the forty-two metapopulations to Harrison's categories is extremely revealing. There are three patchy populations, although they are all too small to be secure (fifteen to twenty-six pairs). Fifteen arrays can be classified as core-satellite, although only six of those have a large enough mainland (core) population to be reasonably extinction resistant. Three metapopulations are of the classic Levins type, and therefore questionably secure, and twenty-one are in the vulnerable nonequilibrium state. So if one looked only at the total species numbers, about four thousand pairs of adults, one might conclude that there were no grounds for alarm.

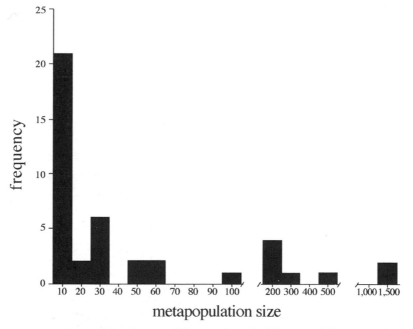

Figure 3.4. Frequency distribution of the number of adult pairs of Florida scrub jays (*Aphelocoma coerulescens*) in the forty-two metapopulations that constituted that endangered species in 1996 (Stith et al.1996, figure 9.9). Note that half of the metapopulations contain ten or fewer adult pairs, and only three have five hundred or more pairs. (Excerpted from "Classification and conservation of metapopulations: A case study of the Florida scrub jay" by Bradley M. Stith et al. Found in Metapopulations and wildlife conservation by Dale R. McCullough, ed. Copyright 1996 by Island Press. Reproduced by permission of Island Press, Washington, DC.)

Placing the birds in these spatially structured metapopulations, on the other hand, reveals that only six metapopulations are reasonably extinction resistant. Moreover, a clear vision is provided to conservationists as to how to develop a species survival plan. This is a wonderful example of the power of the metapopulation approach.

The metapopulation concept represents a major advance for ecology and conservation biology. The parts and processes inherent in metapopulations are summarized in Box 3.1. First, there are patches of habitat suitable for our target species, variously arranged across a landscape. These patches have various properties that are important for our species, including an explicit spatial arrangement. Second, separating the habitat fragments is a matrix, the properties of which are also of critical relevance (this will be discussed in more detail in Chapter 5). The third component is the target species itself, which lives in some or all of the patches of habitat and survives to varying degrees in the matrix, as well. Because metapopulations reference only a single species, they are at the population level rather than the community level of biological organization.

Metapopulation Processes

Besides parts, metapopulations possess characteristic processes. Of central significance is dispersal among the patches. The quantity, quality, and timing of these movements are what give the metapopulation its all-important connectivity. Some movements in and out of metapopulations are also possible, although these must be rare or the metapopulation boundaries would have to be enlarged to accommodate them. The second major process is that of the demography of the target species within the patches. Demographic processes will be influenced by the species' life history features, its morphology, the quality of the patch, stochastic events, and all of the metapopulation properties extending beyond the boundaries of a particular patch. One aspect of demography that will be of particular importance is the synchronicity of demographic behavior across the various demes. Especially relevant in this regard are the timing of population density changes, the rate of emigrant production, and the responses of demes to extrinsic factors such as catastrophes. Finally, metapopulations develop genetic structuring. Variations in genetic makeup across the metapopulation will likely influence the performance of particular demes, their probability of extinction, their rate of production and chances for emigrant success, and potential for adaptability to changing conditions.

Box 3.1.

The Parts and Processes That Characterize Metapopulations

Parts (with their components)
1. Patches of habitat specific for a focal species
   a. spatial arrangement
   b. size and shape
   c. quality
2. Matrix
   a. size and shape
   b. quality for supporting focal species dispersal
   c. spatial variation, including placement and nature of corridors
3. Populations of focal species
   a. inhabiting some or all of the habitat patches
   b. dispersing among habitat patches

Processes (with important influencing factors)
1. Dispersal
   a. timing of dispersal
   b. quality of dispersers
   c. emigration and immigration rates
   d. successful colonization rate
2. Demographic processes and demic extinction
   a. local demic behavior
   b. production of emigrants
   c. acceptance of immigrants
   d. synchronicity among demes
3. Processes affecting genetic structuring
   a. number of demes (occupied patches)
   b. size of demes
   c. rates of emigration and immigration
   d. demic extinction rates
   e. susceptibility to stochastic influences (resilience), including genetic drift
   f. differential selective pressures among demes

All of these metapopulation parts and processes will be discussed in more detail in this and other chapters. Excellent sources for further exploring modern approaches to metapopulation biology include the edited book by Hanski and Gilpin (1997) and *Metapopulation Ecology* by Hanski (1999). A. D. Taylor (1991) discusses limitations on the extent to which species' persistence can be predicted solely on the basis of fragment sizes.

Dispersal

Absolutely critical to the maintenance and long-term prospects of any metapopulation is the dispersal of individuals among the habitat patches. This is what determines the connectivity among the constituent demes and hence is the most important of the vital processes characterizing metapopulations (Box 3.1). Dispersal is rarely a simple process and so needs to be explored in more detail. Corridors can be an important ingredient in this story, because it is essential to understand what conditions will favor dispersal movements by a focal species and thereby improve the connectivity of a metapopulation. Conversely, we may want to know how to control a pest or exotic species, in which case we will need to know how to disrupt metapopulation connectivity.

What Is Dispersal?

The subject of dispersal biology is troubled by much semantic confusion. So we will start with some basic definitions of terms we use in this book (see Stenseth and Lidicker, 1992, for further discussion of semantic issues):

Dispersal—Process of individuals leaving the place where they are resident (home) and looking for a new place to live. This behavior can occur both within and between habitat patches.

Disperser—Individual in the process of dispersal. If the search for a new home is successful, such an individual would be a successful disperser.

Excursion—An exploratory trip away from home including a return home. Such movements may be preliminary to dispersal, and if an excursion ends with the death of the individual, we cannot distinguish it from failed dispersal.

Disseminule—A life history stage adapted for dispersal, for example, planktonic larvae, winged seeds, spores, or spiderlings floating on a spun thread.

Migration—Term often loosely used to refer to any movements of individuals or genes (that is, gene flow), including dispersal. We will restrict the term, as do vertebrate biologists, to mean seasonal movements between breeding and nonbreeding ranges.

Emigrant—A disperser that leaves its home population (deme) and thus represents a loss to that population. Emigration rate is the number of emigrants produced by a deme per unit of time.

Immigrant—A disperser that enters and becomes established in a new population. Immigration rate is the number of new arrivals per unit of time.

Colonist—A disperser that takes up residence in an unoccupied habitat patch. Colonization rate is the number of new colonies established per unit of time.

Phoresy—Dispersal aided by some other species, for example, a deer carrying a seed that sticks to its fur and is dropped in some new location, or a bird eating a fruit and later defecating viable seeds.

Dispersal distance—The distance moved by a disperser from its current home to a new one. The frequency distribution of dispersal distances exhibited by a population (Figure 3.5) is a useful statistic.

Who Disperses?

The first essential point about the dispersal process is that dispersers are not likely to be a random sample of the source population. Dispersal is often biased by sex, age, or genetic makeup. Although there are many exceptions, mammals tend to have male-biased dispersal and birds female biased. Moreover, the sex that dominates among dispersers often tends to travel farther. This implies that if both sexes are needed for successful colonization of empty habitat patches, the process will be constrained by the least dispersive sex. In the extreme case where one sex does not disperse at all, the species will be able to spread only by the incremental addition of new home ranges to the periphery of preexisting ones. For species with this kind of life history, habitat fragmentation poses an almost insurmountable barrier to long-term survival. Every metapopulation would quickly become the nonequilibrium type and be destined for extinction unless connectivity were reestablished.

Dispersal is also often age biased. Often it is young individuals that are most likely to become dispersers. Prereproductives such as recently weaned mammals or recently fledged birds often disperse readily. Sometimes there is a life history stage specifically adapted for dispersal; such stages are called *disseminules*. Among plants, these are seeds or spores.

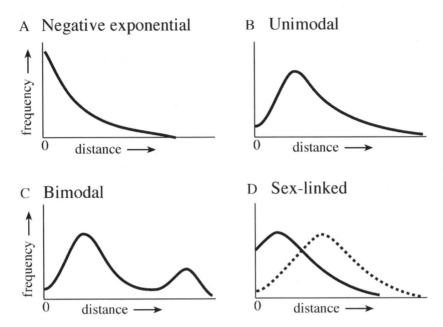

Figure 3.5. Four possible dispersal distance distributions. The negative exponential pattern is characteristic of seed dispersal from a parent plant (seed shadow) and other passively dispersed propagules (A). This pattern is the most frequently used by modelers. Unimodal distributions are usually strongly skewed toward long distance and represent a commonly found pattern in nature (B). Bimodal patterns are found when there is a genetic polymorphism for dispersal behavior, where phoresy occurs, and can result from habitat fragmentation (C). Sex-biased dispersal is well documented in mammals, the males of which often go longer distances, and in birds, the females of which predominate among dispersers (D). When one sex is strongly philopatric, the peak for that sex may be close to zero.

Pollen provides a mechanism for movement of genetic material, but pollen alone cannot establish new colonies. Adults of nonsedentary species also can disperse and often do, but typically that happens in response to environmental cues, such as deteriorating food supplies or the beginning of the breeding season, and not primarily to particular developmental stages.

Although it's a more esoteric bias, dispersers may have significant differences in their genetic composition compared to those that stay at home. This possibility comes from the discovery that some organisms display genetic variation in the tendency to disperse. In such cases, colonizers of empty habitat patches will on average carry more genes encouraging dispersal than other individuals of the same species that stay at home (that is, philopatric conspecifics). The potential evolutionary consequences of fragmentation may well be something that conservationists will be well advised to consider in their planning. As fragmentation begins, we could anticipate selection for increasing rates of dispersal in the metapopulation. If strong social forces initially inhibited emigration, they would likely be weakened in the process. Successful dispersers will be the individuals that establish colonies in empty patches and hence keep the metapopulation viable. However, if fragmentation is progressive, there will be a point where successful dispersal becomes so unlikely that selection will favor those that opt for philopatry and not for accepting the risks of dispersal.

Why Leave Home?

A really intriguing aspect of dispersal biology is the issue of why an individual would want to leave home in the first place. This is a fascinating subject that cuts across the disciplines of physiology, behavior, and ecology, and it is discussed in more detail in Lidicker and Stenseth (1992). The traditional view of dispersal is that it happens when conditions at home become intolerable, thus motivating exodus. Deteriorating conditions can be caused by resource depletion, social intolerance, high levels of predation and/or parasitism, or accumulation of toxic materials. A corollary of this scenario is that those who leave as dispersers are those who tolerate the deteriorating conditions least well. They may be young or old, the sick, or social subordinates that compete least well for mates or resources. Moreover, these difficulties generally occur at times of the year when conditions are most harsh. It follows that the hapless dispersers will face the challenges of travel at times when conditions are poor and they themselves are not in optimal condition for coping with these difficulties. Consequently, they will have little chance of success. This traditional type of dispersal is called *saturation dispersal* (Lidicker 1975).

Saturation dispersal certainly does occur, but now we recognize that much dispersal is of a different sort, namely *presaturation dispersal* (Lidicker 1962, 1975). In this type, individuals leave home for a variety of reasons other than an unhappy home life. In the case of disseminules, there is a morphological imperative. For juveniles, there may be a physiological imperative as they approach sexual maturation. There may be motivation to find unrelated mates so as to avoid inbreeding. At any age, there may be opportunities to colonize empty patches of habitat and have numerous successful progeny. Decisions to move of this type will generally be independent of population density and will most likely occur when environmental conditions are favorable, greatly improving the chances for success. The time course for both kinds of dispersal in relation to density is shown schematically in Figure 3.6.

Whatever the proximal motivation for dispersal, each prospective disperser faces an implicit balance sheet of pros and cons that will influence the likelihood of successful movement. Natural selection will favor those individuals who make the most advantageous "choices" most of the time. Table 3.1 summarizes the potential advantages and disadvantages that may be relevant to any dispersal "decision."

The Demography of Extinction

The second category of metapopulation processes to be considered is that of demographic behavior (Box 3.1). Specifically, we are interested in the probability that the number of individuals in a deme will fall to zero, that is, local extinction. The factors that influence that probability are many and complex (Beissinger and McCullough 2002, Turchin 2003), and this book cannot adequately treat them all. For a succinct overview of the basics, see Lidicker 2002. What we can do is provide a brief outline of the factors that are important in the demography of small populations. We will emphasize those aspects that are most relevant to connectivity among demes, including the role of corridors.

We have emphasized the critical role that dispersal plays in the viability of metapopulations (see also Stacey et al. 1997). At the same time, we have shown that dispersal can be anything but a simple process, making it imperative that conservation planners consider the particulars of dispersal behavior for target species. Moreover, there are other demographic implications of dispersal, both on the local population and metapopulation levels. Movements in and out of populations represent two of the four vital rates that constitute the population growth equation:

$$dN/dt = (b + i) - (d + e),$$

where b, i, d, and e are birth, immigration, death, and emigration rates, respectively. Dispersal movements thus have the potential for significantly affecting demic growth rates (dN/dt). For example, a high rate of emigration could periodically or chronically depress growth in particular habitat patches. Dale (2001) argues that female-biased dispersal in birds will lead to heavily unbalanced sex ratios in small habitat fragments. This in turn

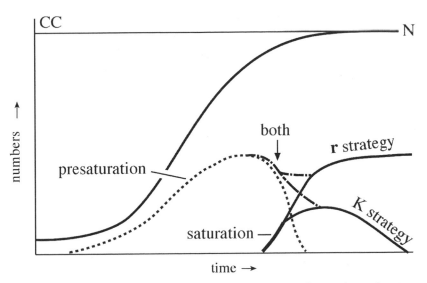

Figure 3.6. The time course of saturation and presaturation dispersal as a function of population density (solid S-shaped line) (Lidicker 1975). CC is estimated carrying capacity for the population; r strategy is a life history mode characterized by high reproductive rates, one corollary of which is that slowing of population growth rates usually occurs by increasing mortality; K strategy is a contrasting mode, in which slow growth is associated with reduced reproduction rather than mortality.

will strongly suppress reproduction in those patches and push the deme toward extinction.

The other aspect of demography that is important here is the synchronicity of demographic behavior among the demes in a metapopulation. If there is much asynchrony among demes, it is unlikely that all of them will go extinct at the same time, and metapopulation persistence is thus encouraged. Much synchrony, on the other hand, could result in a situation where all demes simultaneously suffer catastrophic declines, resulting in simultaneous extinctions and metapopulation disappearance. So it is important to consider what factors encourage asynchrony. At least two things are important in influencing demic synchrony. The first is geographic extent of the metapopulation and amount of variation among the habitat patches. The more spacious a metapopulation and the more variation it encompasses, the less likely it is that a single event, catastrophic or not, will have a similar impact on all of the component demes. Conversely, spatially limited metapopulations with relatively homogeneous conditions throughout will more likely respond to changes in a uniform manner. The

Table 3.1
The Dispersal Balance Sheet

Types of factors	Potential advantages	Potential disadvantages
Environmental	Escape from unfavorable conditions (economic, physical, social)	Uncertainty of finding food, shelter, appropriate social situation
	Reduced exposure to predators and competitors	Greater exposure to predators and competitors
	Reduced exposure to population crashes	
Quantitative Genetic[a]	May find uninhabited or in-completely filled habitat patch	May not find any suitable habitat
	Promiscuity (both sexes have multiple mates in any given breeding season)	Uncertainty of finding a mate
		Strange phenotypes may be avoided
	Frequency-dependent selection may favor rare phenotypes	
Qualitative Genetic[b]	Heterosis and avoidance of inbreeding	Less viable offspring may be produced (breakdown of co-adapted systems; disadvantageous recombinations)
	Greater chance for new and advantageous recombinations	

Source: Modified from Stenseth and Lidicker 1992, Table 1.2.
[a]Quantitative genetic factors are those that influence the quantity of future reproduction by the disperser.
[b]Qualitative genetic factors influence the fitness of future offspring.

second factor influencing synchronization is dispersal. The more the demes are connected by dispersal, the more likely they are to share similar demographic patterns, and hence suffer simultaneous declines. Ironically, while dispersal is essential for metapopulation survival, too much dispersal can be disadvantageous unless metapopulations are large and diverse. Population modeling has suggested that substantial stochastic influences on population dynamics can serve to asynchronize local populations to the extent that the probability of metapopulation extinction is reduced (Allen et al. 1993, Ruxton 1994). On the other hand, Grenfell et al.

(1995) report that if density dynamics are subject to strong seasonal forcing, this can overcome chaotic (stochastic) forces, enhance demic synchrony, and consequently lead to increased likelihood of metapopulation extinction. Kallimanis et al. (2005) support these principles by modeling extinction risk as a function of patch aggregation and the scale of disturbance events. Fontaine and Gonzalez (2005) describe an experimental study of rotifer metapopulations in which both dispersal and temporal environmental variability influenced demographic synchronicity. The strongest synchronicity was produced when high levels of dispersal were combined with periodic environmental fluctuations.

An often overlooked life history trait that can strongly influence a species' dispersal behavior is its social system, which can do that in a number of ways. One of these has already been mentioned, namely that successful colonization of empty patches may require the nearly simultaneous arrival of a large number of conspecifics so that they can establish the necessary social environment for their survival. A corollary to this is the requirement that such coordinated dispersal may mean that groups of individuals will have to emigrate together. The centripetal forces of strong social bonds may make that difficult to organize. Moreover, species that profit from tight social interactions will likely find that dispersal by isolated individuals will be especially precarious. An example of this was reported by W. F. Laurance (1995), who found that the highly social marsupial *Hemibelideus lemuroides* almost never left large patches of tropical rain forest (in Queensland, Australia) to traverse corridors and colonize empty forest patches. Thus, while it remained common in undisturbed rain forest, it was completely absent in fragmented areas. Cockburn (2003) makes the case that cooperatively breeding birds have a reduced capacity for colonization compared to those with the usual single-pair mating system. Finally, social behavior may also discourage immigration into local populations by making it difficult for prospective immigrants to integrate into existing social groups.

Why is it that small populations are subject to very high extinction risks? There are possible genetic reasons for this that will be discussed. But there are also nongenetic factors that may play important roles. These can be random (stochastic) influences or more insidious nonrandom (deterministic) forces. Demographic stochasticity is the process whereby the demographic structure of a small population can be radically altered by random birth, death, emigration, or immigration events. Such alterations may make it difficult or impossible for a small population to survive. For example, a population might suddenly find itself consisting only of males, or only of postreproductive individuals, or subject to some environmental perturbation that temporarily caused all reproduction to fail. Because such changes can happen at random, they are unpredictable in their timing and in their impact. They can leave a small deme either increasingly vulnerable or on the path to extinction.

Obviously, the size and quality of habitat patches will influence the number of the target species that can be supported in the patch. For a given quality, larger patches will usually support larger populations. This is not inevitably the case, however, as larger patches may also support predator and parasite populations that are not present in small patches. Generally, demes occupying larger patches are less likely to go extinct than those living in smaller ones. Even common species may, however, undergo strong seasonal or multiannual fluctuations in abundance. At the low points of these fluctuations, such populations would be subject to a much higher risk of extinction. Planning for species' survival therefore demands that attention be given to low points in the changes in numbers, and not merely to a population's behavior at moderate to high densities. For example, fish populations living in streams subject to Mediterranean climate need especially to be monitored during drought years, when numbers may be at unusually low levels. Related to this last issue is the question of how species respond to stochastic catastrophes. By definition, the occurrence and intensity of such events are unrelated to the state of the local populations. Instead, their causes are to be found outside the boundaries of the subject populations. This means that populations must be able to cope with these catastrophes at any stage in their demographic cycles. If the catastrophe is severe enough to cause extinctions, survival of the metapopulation will depend on it being large enough and/or diverse enough that portions of it will escape the catastrophic impact.

Another class of risks faced by small populations is that of minimum threshold densities caused by antiregulating influences (Lidicker 1978, 2002). This phenomenon is sometimes known as the Allee Effect. Antiregulating forces (sometimes called *inverse density dependent*) are nonrandom, in that their influence is predictably related to population density (Figure 3.7). These effects are the opposite of those caused by the regulating factors normally involved in encouraging population growth at low densities and stopping it at high densities. In fact, they stimulate growth at high densities and discourage or stop it at low densities. By these actions, they are destabilizing influences on population growth (Lidicker

1978, 2002). Such forces are especially well developed in social species but are minimally present in any sexually reproducing species. Research on these forces was pioneered by W. C. Allee (1931, 1951) and F. F. Darling (1938). The latter is especially well known for demonstrating that some colonial species of seabirds cannot breed successfully if their numbers drop below a critical threshold. Another classic example is the inability of a group of muskox (*Ovibos moschatus*) to successfully defend their young against wolf predation if group numbers get too low. Such antiregulating forces also occur in plants, as it has been demonstrated in a variety of species that small patches fail to attract pollinators, leading to reproductive failure and demic extinction (Bawa 1990, Groom 1998, Lamont et al. 1993, Roll et al. 1997). These antiregulating forces generate a minimum threshold density (Figure 3.7). If population numbers fall below that level, the population is destined for extinction, no matter how favorable conditions may be otherwise. In effect, these thresholds "raise the bar" for demic survival by increasing the minimum number of individuals needed for a deme to persist.

A final point about small populations is that human managers can influence metapopulation survival by manipulating connectivity among patches, by controlling deleterious predators or parasites, and by making strategic introductions to either supplement existing demes or establish colonies in empty habitat patches. A good example of a metapopulation in which successful management of this type is a realistic possibility is that of tule elk (*Cervus elaphus nannodes*) in California (McCullough et al. 1996). This subspecies currently exists in twenty-two, mostly isolated, populations. Demes are relatively easily monitored demographically and genetically, providing an excellent opportunity for competent management.

Genetic Structuring

The last metapopulation process to be briefly discussed is that of genetic structuring (Box 3.1). Again, we can provide only an outline for this huge topic, emphasizing some general considerations. Comments will be organized into three topics: 1) the genetics of small populations, 2) genes in a metapopulation context, and 3) evolutionary trends. For more detailed discussion, see Gilpin (1991), Hedrick (1996), and McCauley (1993).

The Genetics of Small Populations

It is to be expected that metapopulations will be composed partly or entirely of populations of relatively small size compared to those in non-

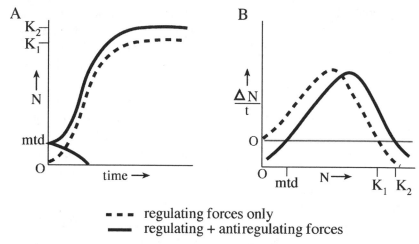

- - - regulating forces only
—— regulating + antiregulating forces

Figure 3.7. The effects of antiregulating factors on population growth rates and equilibrium densities. N equals population size (numbers). Mtd is minimum threshold density, and K_i are equilibrium densities. Graph A shows S-shaped growth with and without antiregulating forces. Graph B shows population growth rate as a function of population size with and without antiregulating influences.

fragmented landscapes. Small size is associated with a number of genetic consequences. These include loss of alleles, increased influence of stochasticity (random changes), and risk of inbreeding depression (discussed further on). If we imagine a population of diploid organisms consisting of ten individuals, it is easy to see why this is the case. Such a population would possess only twenty genes at any particular locus. For those loci that are polymorphic, many alleles present in a large population could easily be missing from this small subsample of genes. From this sampling effect alone, small populations will have reduced genetic variation. The situation could quickly get worse, however, as small populations are subject to significant random changes as one goes from generation to generation. Because of uneven contributions among the ten individuals to subsequent generations, some alleles initially present could readily be lost; in fact, it is quite possible that a single allele will be fixed at a given locus. A single mutation occurring in this population would immediately represent 5 percent of the genes at that locus. Such a mutation could become fixed through random processes even if it were somewhat deleterious relative to the other alleles. If the mutation were significantly deleterious, it would substantially reduce the viability of the deme, since the single individual possessing it would represent a whopping 10 percent of the total population.

Actually, the genetic situation could be a whole lot worse than described in this hypothetical population of ten. That is because the fate of genes from one generation to the next depends not only on stochastic perturbations, but also on what is called the effective population size (N_e). This number is generally smaller than the actual population size (N) and in fact represents the size of a population that would behave genetically the same way as the subject population is actually behaving. N_e is equal to N only if all members of the deme participate equally in reproductive activities for a given breeding season and the population size remains constant. Equal participation requires that all individuals produce the same number of successful progeny, and that the sex ratio be equal. These conditions are all unlikely to be met in any real small population, and to the extent that they are not true, N_e is reduced accordingly. For starters, all nonreproductives, the young, the very old, and those in between that cannot find mates are not counted in N_e. If the sex ratio is not equal among those that do participate in reproduction, N_e is reduced still further. For example, if our population of ten has eight individuals that do the reproducing, but only two of them are males, N_e would be six instead of ten. Beyond this, if there are variations in the reproductive successes among participating adults, N_e is also reduced. Such variations could be due to age and fitness differences, or to chance. Finally, if we are calculating the effective population size over a longer time period than one generation, we must take into account differences in numbers from one generation to another. N_e will be closer to the low points in density fluctuations than to the high values. Of course, all of these factors could be acting simultaneously, so that N_e may be very much less than N.

Last, it needs to be mentioned that small populations risk inbreeding depression, which could further debilitate the deme. Inbreeding depression arises from two causes: (1) increasing chances for deleterious alleles to occur in homozygous form so that the full negative effect of these alleles will be expressed, and (2) loss of fitness caused by the disappearance of heterosis, which is the added fitness (vigor) that often accompanies individuals carrying high levels of heterozygosity in their genomes. Progressive loss of alleles and increased frequency of breeding with closely related individuals both increase the overall homozygosity of small populations. Inbreeding will also lower N_e. As a deme becomes more and more homozygous, it becomes, from a population genetics perspective, increasingly like a haploid organism (one set of chromosomes instead of two and therefore half the total number of genes). That is, N_e approaches $N/2$ from this cause alone.

An interesting and unexpected benefit to genetic diversity within a deme is illustrated by honeybee colonies. Jones et al. (2004) have shown that genetically diverse colonies are better able to maintain stable brood nest temperatures than less diverse colonies. Maintaining optimal nest temperatures over a broad range of ambient temperatures clearly translates into a distinct reproductive advantage. Just how many individuals are needed in a population to maintain long-term fitness is an intensively debated topic. Recently, Reed (2005) has analyzed data from the literature on plant populations and concluded that at least two thousand individuals are required for long-term persistence.

Genes in a Metapopulation Context

When we consider population genetic processes in an array of semi-isolated demes, several important properties are manifest. First of all, the total number of individuals in a metapopulation, the sum of all the demes combined, is ordinarily going to be much less than the size of the population in an unfragmented state. This is simply because fragmentation involves a loss of habitat available to a species since portions of the original habitat are converted to nonhabitat.

The second important fact about metapopulation genetics is that the various demes will tend to become different from one another. This tendency to differentiate is caused by two things: random changes in genetic composition and progressive changes caused by differential selective pressures on the various demes. The greater the spatial extent of the metapopulation and the greater the variation in local conditions across the demic array, the stronger will be the tendency for selective pressures to vary, and hence to influence genetic differentiation within the metapopulation. Random changes (genetic drift) will become increasingly important as deme size decreases, much as small demes experience demographic stochasticity.

Countering the tendency for demes to differentiate is the movement of genetic material among demes (gene flow). Mostly, genes are carried by dispersing individuals, but in some cases gametes can also serve this function, as in the case of pollen. It has been estimated that the arrival of one unrelated and successful immigrant per generation is sufficient to counter the forces of genetic drift, except perhaps in very small populations (Mills and Allendorf 1996, J. Wang 2004). This emphasizes yet another reason conservation biologists need to have detailed information on the dispersal behavior of focal populations.

The final genetic feature to be mentioned is that, in metapopulations, local demes generally have less genetic variation than was available in the original nonfragmented population and that may still be present in the metapopulation as a whole. This, as pointed out, is a consequence of their generally small size and the losses of alleles that go with that, as well as differential selection among the demes. In addition, most empty habitat patches will be colonized by a few individuals, generating a founders' bottleneck and further limiting the initial genetic variation present.

Evolutionary Trends

Generally, land-use planning for conservation is concerned primarily with immediate goals such as saving species from imminent extinction and providing recreational opportunities. It is slowly being appreciated, however, that the special nature of metapopulation genetics generates certain evolutionary tendencies. To the extent that we can anticipate these trends, and plan accordingly, conservation planning will have a brighter future. Except for the vague goal of trying to reduce loss of genetic variability, little attention is typically given to long-range consequences of habitat fragmentation. Partly this is a result of inadequate understanding of the processes involved and their potential interactions, and partly it is simply being aware that predicting the future is precarious indeed. Projections for climate change in the next few decades are forcing us to become much more appreciative of the importance of planning ahead. Here we outline some evolutionary trends that can be associated with increasing loss and fragmentation of natural habitats.

One trend already mentioned is that there will be increased selection pressure for dispersal behavior, at least until fragmentation produces a nonequilibrium type of metapopulation. This tendency may come at the cost of losing strong social forces that may have inhibited dispersal to some extent. Whatever adaptive advantages these social arrangements might have had will therefore be at risk. The good news is that weakening social behavior could have the beneficial effect of reducing the minimum threshold density that often accompanies complex social adaptations. Tilman (1990, 1994) argues that in plants, fragmentation not only selects for improved dispersal but simultaneously reduces competitive ability.

Schmidt and Jensen (2003) studied changes in body size in twenty-five species of Danish mammals over a period of 175 years of increasing habitat fragmentation. They found that small-bodied species tended to get larger and larger species got smaller. These results are consistent with the findings of Adler and Levins (1994), who surveyed data for rodents living on islands and reported that they tended to be larger than their mainland counterparts. Thus, habitat fragmentation may produce evolutionary responses in mammals, and likely other species as well, that are the same as those described for real islands. These so-called island effects are described by Van Valen (1973) and others. If mammals can evolve in this way in seventeen decades, it seems likely that insects and other creatures with higher potential rates of evolutionary change could respond much more rapidly to fragmentation.

There also may be a long-term trend for continuing loss of genetic variation in the metapopulation as a whole. This is because as demes become periodically extinct, there is a loss of all the genetic variation formerly possessed by that population, except as that variation may still be present in successful dispersers from that deme and their progeny. The empty habitat patch will then typically be recolonized by a few founders who carry only a limited sample of the genetic information found in the metapopulation as a whole. Still another trend to be expected is that selection will favor the evolution of stronger inbreeding avoidance mechanisms in order to counter the deleterious effects of too much inbreeding. If this happens, it will help to reduce losses of genetic variation, unless the deme is so small that inbreeding can be avoided only by not breeding at all. Other metapopulation features that would help to reduce genetic losses are large sizes of habitat patches, large numbers of patches, and adequate connectivity among patches. Good connectivity will result in quantitatively more gene flow but also improved quality if there are multiple sources of immigrants to a patch. Spatial heterogeneity in patch conditions will encourage differential selective pressures and hence increase genetic diversity. This is in addition to its beneficial role of discouraging dangerous demographic synchronicity among the demes.

One warning, however, is that if selective pressures diversify the component demes of a metapopulation too much, there is the risk of outbreeding depression occurring. This can happen when individuals who are quite distantly related or adapted to rather different microhabitats mate and produce offspring. These offspring may turn out to be not well adapted to either of the parental habitats, or they may exhibit some other genetic breakdown. Such genetic breakdowns or disruptions of coadapted gene complexes may include balanced polymorphisms, linkage groups, or epistatic complexes (interactions among loci). Sometimes the deleterious effects of outbreeding do not become manifest until the second generation after the initial mating (Edmands 1999). The results of outbreeding may be analogous to what happens in some hybrid zones between species or distinctive subspecies. Decreasing individual fitness, increased parasite loads, and increasing mutation rates have been described in some hybrid zones. If there are a lot of such immigrants they may impede or even reverse a deme's adaptation to local conditions, and thus increase the risk of extinction. Well-meaning wildlife managers have on occasion caused extinction by introducing distantly related individuals to an endangered population and inadvertently hastening extinction. In Pacific salmon, different spawning runs in some river systems represent temporally (instead of spatially) isolated demes. Gharrett and Smoker (1991) have shown that hybridization between even- and odd-year classes of pink salmon (*Onchorhynchus gorbuscha*) leads to poor return rates for the hybrids and their progeny.

Adding complexity, we must contemplate the real possibility that these various genetic risks inherent to small populations will interact with each other and with various environmental stressors. If this should happen, it will be even more difficult to predict the net effects on population viability, as net effects might be less than, equal to, or greater than the sum of the factors acting separately. In an ingenious experiment using gray-tailed voles (*Microtus canicaudus*) living in large outdoor enclosures, Peterson (1996) showed that inbreeding and insecticide application to alfalfa (*Medicago sativa*) had additive negative effects on the voles. Separately, inbreeding and insecticide each negatively affected vole population growth. When both were present, the negative effect was approximately equal to the sum of the two effects acting separately; that is, they were additive.

Finally, it is essential to face the prospect that the reduced overall genetic variation often associated with metapopulation structure will likely make it more difficult to adapt to changing conditions in the future. Roff (2003) suggests, based on the findings of Hoffmann et al. (2003), that tropical rain forest species may be especially at risk in this regard. This impaired capability for evolutionary adaptation may be especially urgent with the current trend toward anthropogenically caused rapid global climate change, which is magnified, of course, by other human assaults on the environment with which organisms must cope, such as pollution. Thus, genetic factors must be considered both in their short-term impacts on demic extinction rates and in the longer-term effects on continuing evolutionary adaptations. We will discuss later how demic connectivity relates to the challenges posed by these trends.

Metacommunity Theory

The notion of metacommunity is a much newer concept than that of *metapopulation*, the term apparently first being used by D. S. Wilson (1992). Like other new ideas, this one has roots in earlier ecological principles. The theory of island biogeography discussed earlier in this chapter is one of those roots. Another is the conservation theory behind the design of parks and reserves. Especially relevant is the rapidly developing subdiscipline of landscape ecology. This last can be traced to the end of the nineteenth century (Hansson 1995, Lidicker 1995b).

So what is a metacommunity? It is an array of patches of a particular type of community variously connected by dispersers. It thus incorporates the metapopulation idea but raises it from the population to the community level of biological complexity. Instead of focusing on a single species of organism, we will now consider all of the species that make up a biotic community. The idea that there are different types of communities requires that we have a classification system so that actual communities can be identified to type. The community type then is the classification category that we use to assign assemblages of organisms living together in specific places into useful groupings. In contrast, the community, or biocenosis, is an actual place on the earth, along with the particular assembly of organisms living there. These two ideas are often confused, but it is critically important that we keep the distinction clear. There is much controversy about what constitutes the most useful or perhaps "natural" definition for community types, but there is no doubt that classifications are possible and useful for communities. It is also plausible that different classification systems may be used for different purposes. In practice, we often use "habitat" to be loosely equivalent to some community type. This is justified because often numerous species of organisms may find a particular community type to be approximately equivalent to their habitat. Guichard et al. (2004) discuss the metacommunity concept as applied to marine communities.

It may be helpful to note that we can recognize four types of metacommunities based on the underlying primary reason for the disjointed distribution of a community type over space; only one of these is based on human modification of the environment:

- *Heterogeneous physical environment.* The physical environment is quite variable over the earth, and organisms respond to this by living only in those places with characteristics conducive to their survival and reproduction. Variables such as latitude, altitude, and topography (slope, aspect) help to define the mix of creatures living in a particular spot. Also the substrate, or local physical context, is critically important. In terrestrial environments, this can be soil type and depth. In aquatic environments, water depth, salinity, flow rate, pH, oxygen content, temperature, and nutrient content are all important, as is the nature of the bottom. Disjunct distributions of these variables result in a fragmented distribution of community types, however they might be delimited.
- *Biotic heterogeneity.* Even in a given physical environmental setting, the organisms present may vary because the extent of the suitable site may be too small for some species to live there, or it may be so isolated that some species cannot reach the site, or there may be barriers like oceans, mountain ranges, or large rivers that constrain the mix of species present on the site. Finally, physically similar sites may have different communities inhabiting them because some may be recovering through succession from natural catastrophes.
- *Island archipelagos.* As discussed earlier, oceanic and continental islands may hold different communities because of differences in their age, size, distance from sources of colonization, and chance.
- *Anthropogenic fragmentation.* An increasingly large number of metacommunities are the direct result of humans destroying one community type and establishing another in its place—for example, agricultural crops, suburbia, clear-cut forests, polluted rivers.

Next, we can ask if metacommunities have characteristic parts, as do metapopulations (Box 3.1). In fact, they do, and they are much the same as for metapopulations. There is the physical place of the patch with its various attributes, and there is the matrix of different community types that make up the intervening space between the fragments. However, the living component is not merely a focal species but the entire assembly of species that give the community the properties of its community type. Not all of the community fragments will have an identical list of component species, but they will all have a similar collection of species and particularly will have those species that help to define the community type. These last will be those with high dominance (influence on the nature of the community) or that are particularly abundant or that are diagnostic for some reason for this particular community of organisms. The mix of species may include nonnatives or exotics, as well as species that are part of the community only seasonally. As we will discuss in Chapter 8, characterizing communities and assigning them to types are an important aspect of corridor site selection and ecologically based land management generally.

Metacommunity processes are summarized in Box 3.2. This list does not include the many processes carried out by communities that are independent of their location in a metacommunity, such as primary productivity. Just as with metapopulations, patches are variously connected by movement of individuals, they suffer extinctions, and new patches can be established. And, of course, the entire metacommunity can decline to extinction or suddenly be eliminated by a catastrophe, anthropogenic or natural. In an experimental study of simple metacommunities involving only a species of bacteria, a flagellate, and a ciliate, it was demonstrated that smaller and more fragmented communities went extinct faster (Burkey 1997). The species richness of patches may be highest at intermediate levels of connectivity (Kneitel and Miller 2003), as strongly isolated patches will suffer higher extinction rates whereas extensive connectivity will likely ensure that dominant competitors and predators will reach all patches and thereby suppress diversity.

Loss of a trophic level in a patch will usually change the demographic behavior of the remaining species, so that a cascade of further extinctions may result. The number of trophic levels that a patch can support depends on the interaction of patch quality and size (Figure 3.8). Patch quality measures the availability of food and other resources and is often strongly correlated with primary productivity of the patch. For a given patch quality, the number of trophic levels possible will increase with patch size. Changing quality, however, will affect the minimum size needed for support of an additional level. Added trophic levels can include parasites as well as predators. Patch isolation could also be a factor in the absence of higher trophic levels. A strongly isolated patch such as an oceanic island or isolated mountaintop may not be detected by dispersing predators or parasites. In the epidemiological literature, the minimum size of a host patch that can support a disease parasite is called the *critical community size* (Grenfell and Harwood 1997). Below that size, the disease is present only intermittently or not at all. In the well-studied case of the measles virus infections in humans, host populations of 250,000 or fewer escape

Some Important Metacommunity Processes and Factors That Affect Them

1. Dispersal
   a. species' varying ability and inclination to move among fragments
   b. timing of species movements
   c. new community (patch) establishment
   d. edge effects
2. Fragment dynamics
   a. extinction
   b. seasonal changes
   c. succession
   d. number of trophic levels
3. Ecosystem services to humans

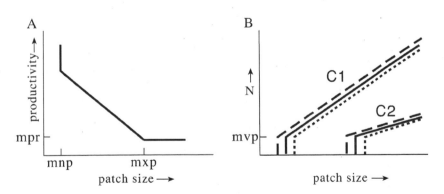

Figure 3.8. Graph A shows the interaction of habitat patch size and quality (productivity) on its suitability for a target species. mpr is the minimum level of patch productivity that can support the target species; mnp is the minimum patch size that can sustain the target species regardless of the productivity level; mxp is the smallest size of patch that can support the target species at mpr. Graph B shows the effect of patch size and quality on the number of trophic levels that can be supported. C1 and C2 are successive consumer trophic levels, N is size of the population, and mvp is the minimum viable population size. Each curve is shown with dashed, solid, and dotted lines representing good, medium, and poor quality, respectively, for the focal patch. (Figure from Lidicker 2000.)

persistent outbreaks of the disease (Grenfell and Harwood 1997, Keeling and Grenfell 1997).

Communities, like populations, often exhibit seasonal or longer-term fluctuations. Such changes need to be accommodated by fragments so that the species they harbor can persist. Otherwise, even species that are often abundant can be lost during seasonal or multiannual periods of low numbers. Another seasonal phenomenon that may be important in community dynamics is the periodic arrival and departure of migrants or overwintering species. Less easy to accommodate are stochastically occurring catastrophes. Ideally, a patch needs to be large enough that portions of the patch will survive such devastating events, and the effects will be reversible.

One path to patch extinction is the gradual change in community structure through the loss of species and perhaps the gain of new ones invading from the surrounding matrix. Eventually, the community may become so modified that it can no longer be considered representative of the original community type. Such altered communities can sometimes be restored to their original type by succession, by gradual recolonization of characteristic species that were lost, or by human intervention. Whether that is possible will depend not only on the size and quality of the patch, but on the degree of connectedness to other patches of the same sort. Species losses from habitat patches are not random. Large species, those on higher trophic levels, and less vagile species will tend to be lost first. Moreover, corridors connecting patches will filter out those species for which the corridors are inaccessible or unsuitable. The role of corridors in metacommunity maintenance will be explored further in Chapters 5 and 6.

Managers and conservationists should also keep in mind that natural or nearly natural community types provide so-called ecosystem services that benefit all life, including humans (see Chapter 1). Processes such as carbon dioxide sequestration, production of oxygen, cleaning and storage of water, soil formation, decomposition of wastes, breakdown of toxins, harboring of pollinator species, control of agricultural pests, maintenance of biodiversity, stabilization of climate, and many more (Daily 1997) are increasingly critical as humans continue to degrade their home planet. These functions may also be provided by anthropogenically generated community types but often less efficiently and with concomitant loss of biodiversity.

As mentioned in Chapter 2, one of the long-standing debates in conservation biology is on the question of whether community types are more securely conserved with single large reserves or several small ones (see Box 2.1). This discussion generally is couched in the context of the design of parks and nature reserves, which usually contain many community types. However, it can also apply to metacommunity viability, as the same issues are relevant. Already we can see that there is no simple answer, and that much depends on how large is large, how small is small, how many small patches there are, the distribution of patch sizes, and how effectively the patches are connected. If community patches are severely isolated, maintenance of their original composition may require that they be quite large (Figure 3.2). Rolstad (1991) estimates that isolated spruce forests in northern Finland need to be at least 20,000 hectares (49,400 acres) to retain their original passerine bird assemblages. Retention of nonpasserines, such as grouse, may require even larger reserves, perhaps 100,000 hectares (247,000 acres). If a reserve targets species of special conservation concern, its design can be tailored to favor the particular requirements of those species. One design will not likely be optimal for all the species in the community. In all cases, the large patches, or alternatively the array of fragments, need to be large enough to withstand the occasional but unpredictable catastrophes that are unquestionably part of life on this planet. We will revisit this controversy in later chapters.

Key Connectivity Considerations

In Part II, we review the general concepts of corridors that could apply to a wide variety of landscapes, and point out that each project will have a variety of site-specific factors that need to be considered. Insights from landscape ecology reveal that what surrounds a potential corridor must be considered, in addition to the properties of the corridor itself. We focus not only on the benefits of corridors but also on the potential problems that can result from various methods used to increase connectivity. We place the range of approaches as well as their caveats into context to make them easier to assimilate and incorporate into real projects.

As a wide variety of approaches are employed to enhance connectivity across human-altered landscapes, we begin in Chapter 4 with an overview of those approaches and their intended benefits. We describe various types of landscape linkages that may serve to enhance connectivity, from greenbelts to wildlife passageways to hedgerows. We also review the variety of scales for which corridors are pertinent, from highway underpasses to continental corridors, and we explain some of the benefits that corridors provide to biodiversity conservation and human quality of life. In Chapter 5, we then focus on the importance of the landscape context, which influences biodiversity, and the conservation tools that might be most appropriate. We discuss the importance of the landscape elements surrounding core habitat and corridors, also referred to as the matrix. The matrix may directly influence connectivity, and it may affect the utility of corridors that pass through it. Unfortunately, there is a limited amount of information on how the mosaic of habitats across the landscape can affect biodiversity patterns and ecosystem processes. A number of factors, ranging from edge effects and invasions of nonnative species to behavioral attributes, could reduce corridor effectiveness and even further imperil a species. Therefore, in Chapter 6, we explain that corridor conservation, while generally viewed as a benefit to the existing problems of fragmentation, may also have some drawbacks. That chapter concludes with important cautionary points that need to be recognized when relying on corridors to increase connectivity. Being aware of the potential pitfalls may help us to avoid them.

CHAPTER 4

Approaches to Achieving Habitat Connectivity

Habitat fragmentation resulting from increasing human activities in natural areas poses a great threat to the long-term conservation of biodiversity worldwide, as discussed in earlier chapters. Corridors are important because they may be a tool for maintaining viable populations of biota in fragmenting landscapes by enhancing connectivity (Forman 1995, Kubeš 1996, A. F. Bennett 1999, Perault and Lomolino 2000). In recent years, there has been increasing research on the subjects of connectivity and corridors. Because corridors are increasingly being implemented worldwide, it is important that we use our knowledge from the research to date to direct our conservation efforts.

The purpose of this chapter is to examine methods for achieving connectivity among habitat fragments by synthesizing corridor research and on-the-ground conservation examples. To begin, we discuss the broad use of the terms *connectivity* and *corridor* and introduce landscape elements that may function as corridors. We then review the different types of corridors that may improve connectivity and discuss the applicability of the corridor concept at various spatial scales. Finally, we introduce some of the known and theoretical benefits of corridors, both to conservation of biodiversity and to human quality of life.

What Is a Corridor?

A number of definitions of corridors and connectivity have been proposed over time. In earlier references, corridors were defined as routes that enhanced speedy and unselective spread of biota between regions (Perault and Lomolino 2000). The Ninth U.S. Circuit Court of Appeals defined corridors as "avenues along which wide-ranging animals can travel, plants can propagate, genetic interchange can occur, populations can move in response to environmental changes and natural disasters, and threatened species can be replenished from other areas" (R. Walker and Craighead 1997). Others have described corridors as linear landscape elements that connect two or more patches of natural habitat and function to facilitate movement (Soulé and Gilpin 1991). Connectivity has been said to describe the extent to which flora and fauna can move among patches, rather than the linear landscape element described as a corridor (Hansson 1995; Figure 4.1). Many European countries developed their own legal definitions of corridors, emphasizing different objectives and approaches to biodiversity conservation, and these are part of the European Union's effort to design and implement regional ecological networks (Jongman 2004). Sometimes corridors are referred to as habitat corridors, wildlife corridors, or ecological structures. They can be part of or the same as ecological or habitat networks, which encompass core areas, corridors, and connecting nodes; or they can be synonymous with greenways, greenbelts, or open space.

In this book, we use *connectivity* to refer to the extent to which a species or population can move among landscape elements in a mosaic of habitat types (Chapter 3; P. D. Taylor et al. 1993, Tischendorf and Fahrig 2000). This necessitates linkages among individuals, species, communities, and ecosystems at appropriate spatial and temporal scales (Noss 1991). Corridors are one means of achieving connectivity. In Chapter 3, we defined corridors as any space identifiable by species using it that facilitates the movement of animals or plants over time between two or more patches of otherwise disjunct habitat (Lidicker 1999). Such movement may occur in a matter of minutes, hours, or over multiple generations of a species. Corridors may encompass altered or natural areas of vegetation and provide connectivity that allows biota to spread or move among habitat fragments through areas otherwise devoid of preferred habitat (Andreassen, Ims, and Steinset 1996, Perault and Lomolino 2000). Landscape elements that function as corridors may also serve multiple other purposes, providing aesthetic amenities, ecosystem service values, cultural heritage protection, and recreational opportunities (Ahern 1995, Fábos 2004). Some landscape elements are unintentional corridors, providing connectivity for biota without being designated for that purpose.

Corridors can be viewed over broad spatial and temporal scales. At one extreme, we have corridors that connect continents, including the Isth-

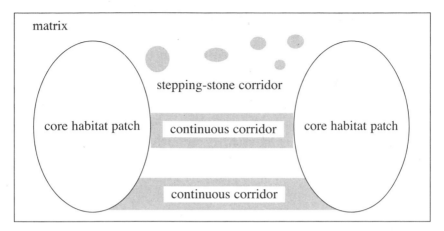

Figure 4.1. A graphical representation of two core habitat patches connected by two continuous corridors and a stepping-stone corridor, all surrounded by matrix.

mus of Panama, and land bridges such as the Bering Strait that appear and disappear with sea level changes over time scales of millions of years. On a subcontinental scale are efforts to connect forest communities from southern Mexico into Panama (Kaiser 2001) and a grand plan to connect the Yellowstone area in Wyoming north through the Rocky Mountain chain to Yukon and Alaska. (the Y2Y project, described in more detail later in the chapter). A successful corridor on a regional scale is the Braulio Carrillo National Park in Costa Rica, which was established to connect Atlantic lowland rain forest (La Selva Biological Station) with high-altitude protected areas. Most land planning efforts that involve issues of connectivity, however, are at a more modest scale, from less than one to a few hundred kilometers. Even there, the temporal scale may be critically important, as the impact of a corridor or lack thereof may be slow to become evident. Lindborg and Eriksson (2004) found that species richness in fragments of seminatural grasslands in Sweden reflected patterns of connectivity one hundred years ago rather than the current or recent landscape configuration. This example illustrates that fragmented communities take time to equilibrate to current conditions, generating an extinction debt or lag effect.

The Donaghy's Corridor project in Australia exemplifies the more typical scale of corridor projects. Here a 1.2-kilometer (.74-mile) corridor attempts to link the World Heritage reserves of the Lake Barrine section of Crater Lakes National Park with the Gadgarra State Forest in the tropical rain forest part of Queensland (Tucker 2000). Over one hundred species of rain forest plants were established over four years from local seed sources along Toohey Creek, a waterway connecting the two protected areas that run across John Donaghy's property. Local volunteers, private landowners, and government agencies worked together to maximize planting success. The planted corridor is now protected with a conservation agreement that establishes the area as a nature refuge and excludes livestock. The corridor will be monitored and maintained to ensure that it continues to preserve diverse vegetation and attract dispersing animals.

At even smaller scales, a corridor may be a structure that crosses a road or canal, or a matrix only tens of meters or yards wide. Small-scale corridors might be trails or paths that guide organisms through thick vegetation or over difficult topographies. Even odor trails, pathways marked or scented by wildlife, such as those established by mammals and ants could qualify as corridors (Kozakiewicz and Szacki 1995, Liro and Szacki 1994). They are linear, enhance the movement of organisms, and may connect habitat patches. Although conservation planners generally deal with medium-scale corridors, functionally meaningful corridors may transcend a broad range of spatial and temporal scales.

Types of Corridors

Different kinds of landscape elements enhance connectivity. Many elements serve as corridors that are not explicitly designed for the purpose, such as roadside vegetation, fencerows, and greenways. In other cases, corridors are purposely retained, maintained, and restored to facilitate landscape connectivity for individual species, groups of species, or entire ecological communities. Here we emphasize the potential of landscape elements to enhance connectivity. In later chapters, especially Chapters 6 and 7, we will address some factors that make these landscape elements more or less effective and sometimes even detrimental to native species.

Unplanned Corridors

Landscape elements that enhance connectivity but exist for other reasons are de facto corridors. These are often locations where optimal habitat or even marginal habitat is left undisturbed, providing a different vegetative structure from the surrounding matrix (adjacent land uses and habitats) (Kubeš 1996). In highly modified environments, such remnant habitat may be disturbed, invaded by exotics, or sparsely vegetated, but some plants and animals may still be able to disperse through it or survive within it. Fencerows, windbreaks, roadside vegetation, and ditches may serve to enhance connectivity (A. F. Bennett 1990, Merriam and Lanoue 1990, Wegner and Merriam 1990, Crome et al. 1994, Kubeš 1996, Joyce et al. 1999, Poague et al. 2000).

Roadside corridors (vegetation strips along roads) are an example of de facto habitat that can have both positive and negative effects on connectivity for native biota. They can offer habitat to both plants and animals and can act as a conduit for movement among habitat patches. The presence of indigenous vegetation should enhance the ability of roadside corridors to act both as a conduit and as supplementary habitat (A. F. Bennett 1991). One example where roadsides provide connectivity is in Southern California, where revegetated highway rights-of-way enhance connectivity for native rodents and urban-adapted birds between habitat patches, although more sensitive bird species do not use them (Bolger et al. 2001). Downsides to roadside corridors may include not serving specialist species and drawing species to roads, where they may ultimately be killed by automobiles.

In agricultural landscapes, fencerows, unmanaged ditches, creeks, and shelterbelts can all serve as de facto corridors. Often native or nonnative vegetation along fences is not managed, offering vertical vegetative structure that some species of plants and animals use to live in or travel through. One experimental study, for example, showed that mice (*Peromyscus leupocus*) preferred structurally complex fencerows over other landscape elements (Merriam and Lanoue 1990). Similarly, vegetation along ditches and creeks is often left and can serve as both habitat and a conduit for species traveling among larger habitat patches. In northern California, detection of mammalian predators was elevenfold higher along creeks than in vineyards (Hilty and Merenlender 2004), and bird and small mammal diversity was higher along structurally complex creeks than in denuded riparian areas or vineyards (Hilty 2001). Shelterbelts, tree rows planted to prevent soil drift and hold snow on fields, are another element in agricultural landscapes used by some species of wildlife. For example, most movements of studied migratory bird species that breed in agricultural shelterbelts in North Dakota were found to occur in shelterbelts and between connected rather than unconnected sites (Haas 1995). These studies all suggest that vegetation structures within agricultural landscapes can function as movement corridors and even provide habitat for some species of wildlife. These linear elements can also be problematic for native wildlife, however, by inhibiting movement of some species and harboring or boosting the presence in the landscape of exotic species and predators that might not otherwise be able to persist.

Identifying landscape elements that may already, or could with some restoration, serve as corridors is important in planning species' conservation. Such landscape elements can be assessed, as discussed in Part III of this book. Variables such as dimension, vegetative structure, and overall landscape context will affect their utility (Chapter 7).

For the most part, the landscape elements described here are most likely to facilitate generalist species and may not serve specialists; they could also cause a net loss for some species and result in mortalities (Chapter 6). Still, in heavily impacted landscapes where setting aside or restoring larger corridors is not feasible, enhancement of de facto corridors may help retain what species do remain in the landscape. Because most evidence of de facto corridor use is based on short-term observational studies, future evaluation and monitoring will be important to refine our understanding of their utility.

Planned Corridors

Even when corridors are planned, connectivity for biodiversity may be only one of the purposes. For example, greenways, also referred to as open-space systems or greenbelts, can potentially provide connectivity (Ahern 1995). These are areas that are set aside for recreation, culture, and ecosystem services, usually within a densely developed landscape and often in cities, suburbs, and adjacent countryside. They can include everything from natural habitat to farmland to areas unfit for development because of susceptibility to floods, topography, or other factors (Zube 1995, Arendt 2004).

Many point to Frederick Law Olmsted as the founder of the greenway concept. In his planning of the Boston Emerald Necklace, or Boston's park system, in the early 1900s, he used greenways to link parks through the city (Fábos 2004). The 1987 report of the U.S. President's Commission on the American Outdoors launched today's greenway movement by describing a future vision of greenways that offer people access to open spaces near where they live, as well as linkages to rural and urban open spaces (Fábos 2004).

Current greenway planning around the world generally focuses on multiple goals, one of which is sometimes biodiversity conservation. Often, efforts focus primarily on control of urban expansion and on recreational benefits (Ahern 1995, Kubeš 1996). Other objectives can include water resource planning or historical or cultural resource protection (Ahern 1995). Attainment of connectivity for biodiversity in the context of these other goals depends on site-specific variables and the species of focus. In some cases, human activities may inhibit the corridor function for species (Haight et al. 2005). Trampling of vegetation, purposeful or inadvertent introduction of nonnative species, and wildlife harassment by pets are some examples of factors that may impair a greenway's connectivity value for biodiversity. Even so, greenways not explicitly focusing on conservation should be evaluated for their potential as corridors by providing habitat, acting as conduits, and even harboring source populations.

Examples of greenways abound. Georgia is focused on accommodating human interest as well as natural resource conservation through greenways (Ahern 1995, Dawson 1995). A study of greenway potential across the state balanced natural and human values. The former included natural resources, environmental quality, and aesthetics, while the latter included human use, accessibility, market demand, and land use (Dawson 1995). This study has served as the basis for land and access rights acquisition. Carefully planning the uses of the purchased lands would help ensure that these greenways could maintain connectivity value for natural communities.

An urban example of greenways is the East Bay Area Regional Parks in northern California, which are situated directly adjacent to the highly urbanized Bay Area. Figure 4.2 illustrates the hard edge between the natural habitat and the human-occupied parts of the landscape. The greenways offer recreational opportunities and some benefits for biodiversity conservation. One problem with these small urban natural areas is that they are often ecologically out of balance. For example, they usually lack the large predators that previously inhabited these areas. As a consequence, these parks and surrounding urban areas often are plagued by overabundant populations of prey, such as deer, which increase due to both the lack of predators and the lush gardens along the urban fringe (McShea et al. 1997). The spread of exotic species and fire control are also common concerns that must be managed in such urban parks. Domestic cattle and goat grazing is sometimes employed in the East Bay Parks and other urban greenways for fire control, which can further compromise biodiversity conservation. Such management can impair the ability of these greenways to serve as habitat or conduits for species' movements. While some species may thrive near high-density human areas and in well-used greenways, careful planning is needed to retain more sensitive species.

Figure 4.2. The abrupt edge between the Bay Area Regional Parks and the cities of Oakland and Berkeley, California. (Photo by David House, 2005.)

For more examples of greenways and their diverse purposes, two special sections of the journal *Landscape and Urban Planning* include multiple contributions about greenways planning. The first, "Ecological Resources and Nature Protection in Greenway Planning," was printed in volume 33 in 1995; the second, "International Greenway Planning: An Introduction," appeared in volume 68 in 2004. These articles emphasize the need for greenway planning to include landscape-level considerations, including promoting conservation by using design guidelines developed with sound ecological principles such as those discussed in Chapters 3 and 8. In general, the smaller, more heavily used and less biologically intact greenways will likely have less biodiversity value compared to larger, more intact greenways with less human activity. However, few studies provide concrete evidence of greenways facilitating movement of genes or species. Also, most greenway planning is occurring in areas where species sensitive to human disturbance have already disappeared, so that the greenways are likely serving more human-adapted and generalist species.

Some corridors focus solely and explicitly on ecological needs. They may buffer linear landscape elements of particular importance to biodiversity, such as riparian zones; conserve priority areas for individual species conservation; or promote community integrity across broad regions. In contrast to de facto corridors, these are often designed using scientific principles, biological surveys, and models to help determine landscape location. The major assumption in designating such corridors is that they will enhance conservation by promoting one or more connectivity goals (Box 4.1).

Buffering Riparian Zones

Arguably one of the most important landscape elements for biodiversity is the riparian corridor. Riparian corridors are made up of vegetation growing adjacent to creeks and rivers that are sometimes retained in human-dominated landscapes (Figure 4.3). Riparian areas support a disproportion-

Box 4.1.

Planning a Corridor: Biodiversity, Scale, and Goals

The following lists contain the hierarchical levels of biodiversity commonly considered when planning a corridor, the scales at which corridors are implemented, and the potential goals that can result from corridor implementation.

Levels of Biodiversity
    Individual (of a species)
    Deme (of a species)
    Species
    Community
    Landscape

Spatial Scale (of Linkage)
    Local (e.g., underpass)
    Regional (e.g., river corridor)
    Continental or cross-continental (e.g., mountain range)

Potential Goals
    Daily movement (e.g., access to daily resources)
    Seasonal movement (e.g., migration)
    Dispersal (e.g., genetic exchange, mate finding)
    Habitat (e.g., wide greenway corridor)
    Long-term species persistence (e.g., adaptation to global warming)

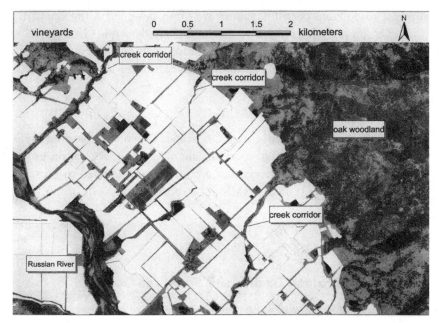

Figure 4.3. Creek corridors composed of remnant riparian vegetation in Alexander Valley's oak woodland and vineyard landscape (Sonoma County, California). To emphasize the vineyard matrix, a map compiled in 1997 of vineyard blocks (white) has been superimposed onto the ortho photo.

ately large amount of biodiversity compared to other landscape elements, and conserving these sites can provide multiple natural resource benefits (Harris et al. 1996). Vegetation strips along river systems also protect in-stream biota by controlling erosion and providing shade to keep water temperatures cool. Retaining buffers along streams can benefit terrestrial biota as well. For example, bird species richness and abundance appear to be greater where adequate riparian buffers are retained, according to studies of forests ranging from boreal forests in Sweden to riparian forests in California and Georgia (Hodges and Krementz 1996, Hilty 2001, Hylander et al. 2004). Buffer zones around wetlands and riparian habitats also have been found to be important for amphibian and reptile populations (Semlitsch and Bodie 2003). Riparian buffers around lakes, rivers, and wetlands may be explicitly retained for conserving species, or they may be de facto, the result of policies such as those oriented toward water quality enhancement.

Establishing and preserving vibrant riparian corridors is a good approach to conserving species, but conservation of riparian corridors alone is inadequate. First, landscape context is important, and corridors within a less intact landscape will be less effective. Second, creek and river corridors can lead wildlife into areas of human activity instead of to other habitat patches. When this occurs, corridors are essentially dead ends. In Bozeman, Montana, black bears (*Ursus americanus*) follow creeks into town each year, only to discover that creeks are eventually funneled underground or that their riparian vegetation has been stripped away. The bears suddenly find themselves in human neighborhoods and often become disoriented or begin eating trash (Haines 2004, McMillion 2004).

Corridors for Individual Species Conservation

In addition to protecting specific landscape elements such as riparian areas, corridors may be mandated in individual species' management plans. For example, dispersal corridors were proposed and identified through logging areas for spotted owls (*Strix occidentalis caurina*) in the U.S. Northwest (USDA/USDI 1994). Similarly, an important part of the recovery of panthers (*Puma concolor coryi*) in Florida has been to identify and create safe corridors for them to move among remaining habitat fragments (Cramer and Portier 2001). This has involved both selecting specific locations such as road underpasses and conducting broad-scale landscape connectivity analyses. Road underpasses along Florida's Interstate 75, also known as Alligator Alley, enhance connectivity and reduce panther deaths on roads because fencing inhibits road crossing and guides them to the underpasses. The Pinhook Swamp was identified as a regional link connecting the Osceola National Forest to the Okefenokee National Wildlife Refuge for panthers as well as black bears (Harris et al. 1996).

Corridors That Enhance Community Integrity

Connectivity may be promoted to protect biotic community integrity or suites of species moving among parks or protected areas across large regions (Ndubisi et al. 1995, R. Walker and Craighead 1997, Noss 1991). Corridors may be part of an ecological network that includes multiple core protected areas and corridors where connectivity minimizes the isolation of protected areas, such as in national parks. Scientists used modeling techniques to identify links, for example, between Glacier National Park in Montana; Yellowstone National Park, mostly in Wyoming; and the Salmon Selway Ecosystem in Idaho that would benefit grizzly bear, elk, and cougar (*Puma concolor*) (e.g., Mietz 1994, R. Walker and Craighead 1997). Another coarse-scale corridor is the Kibale Forest Corridor in Uganda, which connects two parks, Queen Elizabeth and Kibale National Parks (Baranga 1991). The long-term integrity of both these corridor projects is a challenge, as many people reside within the corridor boundaries and alter the natural habitat. However, these corridors may function for some species, especially if targeted restoration efforts are made and the resident human communities are engaged.

Whether designating a planned or an unplanned corridor to help conserve a species or a community, the purpose is to enhance connectivity (Box 4.1). What can serve as connectivity for one species may act as a barrier to movement for other species. To avoid unintended negative consequences, it is important to evaluate how a corridor may affect all native species in the landscape. Later in this chapter, we discuss benefits of corridors to conservation of biological diversity, and in Chapter 6 we discuss potential negative consequences.

Goals of Corridors

As background to discussing goals that corridor projects seek to achieve, we briefly review a few points about corridors that often cause confusion. First, corridors can target some or all levels of biodiversity (Box 4.1). Second, they occur at many different spatial scales. For example, some corridors may be a few meters or yards long to facilitate movement of smaller species, while others may span one or several countries to provide a conduit for biotic movements over a long time period (T. W. Norton and Nix 1991). As will be discussed in Chapter 7, researchers and land-use planners must be explicit about the scale of a proposed corridor and the species that it is designed to benefit. Third, corridors may provide connectivity for one species and not another due to species' different operational scales and habitat requirements. Finally, because the integrity of a community may affect individual species' survival, connectivity planning for entire communities should be considered where possible, rather than focusing on individual species (Lidicker and Koenig 1996). With the above issues addressed, corridors may achieve any of five connectivity objectives: daily movements, dispersal, seasonal movements, habitat connectivity, and long-term species persistence (Box 4.1).

Species, demes (local populations), and individuals often move from one patch to another each day. One common reason for daily trips is to access resources such as water. Corridors can ensure that travel routes to necessary resources are not severed to avert imperiling a population. Corridors may also protect their key habitats or help animals avoid predation they might suffer in crossing modified habitat or human-dominated landscapes (Noss 1987). Creek setbacks, for example, provide habitat for otters (*Lontra [=Lutra] canadensis*) and allow them to move along stream courses. Corridors can allow individual bears to move around a large home range (Jordan 2000). Box 4.2 describes the plight of an individual cougar to survive in a human-dominated landscape and the effort to conserve a corridor that connects patches of habitat used by that individual.

Enhancing survivorship of dispersers is another common goal and can help to conserve individuals, demes, and species. Dispersal among relatively isolated populations in a metapopulation is important for preserving genetic diversity and reestablishing populations in habitat fragments where a species has become locally extinct (Chapter 3; McCullough 1996). Given increasing evidence of huge distances traveled by dispersing individuals (e.g., Inman et al. 2004) and generally low survivorship in human-dominated landscapes (e.g., Beier 1995), understanding how organisms move across the landscape will help in conserving key landscape elements and potential travel pathways. Unfortunately, few studies have adequately documented dispersal events, although global positioning systems (GPS) are making such studies more feasible (Lambeck 1997). A Greater Yellowstone Ecosystem study using GPS collars for wolverines (*Gulo luscus*), a species that exists naturally at extremely low densities, found that dispersing individuals may move hundreds of kilometers or miles when looking for mates or establishing a new territory (Inman et al. 2004; Figure 4.4). Their preferred habitat in the contiguous United States appears to be high boreal habitat, which is found in discontinuous mountain ranges. Sustaining a viable wolverine population in the Greater Yellowstone Ecosystem requires understanding how wolverines perceive landscape connectivity, and in particular how and whether they will move through increasingly human-occupied valleys between preferred mountain habitats. If they avoid areas of human activity, viable dispersal corridors may be critical in areas with increasing human development to ensure that their preferred mountain habitats do not become increasingly isolated.

In addition to corridors being important for dispersal, some individuals, demes, and species require corridor conservation to facilitate annual or seasonal migrations. For example, pronghorn antelope (*Antilocapra americana*) migrate up to 270 kilometers (167 miles) between Grand

Box 4.2.

Travel Route Perils of an Individual Mountain Lion

The following tale is excepted from Royte (2002):

The mountain lion was a healthy male, and young. He was born in the Santa Ana Mountains of southern California, probably in the dry, rugged hills near the seaside town of San Clemente. . . . At 18 months, the lion—known as M6 to the scientist who tracked his movements—began to roam farther, looking for a home range of his own and a mate.

One night M6 headed north. At midday he rested; when darkness fell, he resumed his trek. About 80 km (50 mi) into his journey, he left the conifers of the higher peaks in the Cleveland National Forest and dropped down into the sage scrub of Coal Canyon. Its stony creek bed led him into a broad, sandy outwash. Here M6 took stock of his predicament. An eight-lane freeway, Highway 91, the major thoroughfare from Riverside County to Los Angeles, blocked his progress. Hundreds of cars every hour streamed past. M6 sniffed out a derelict underpass. . . . For 187 days M6 stayed put, patrolling 12,000 acres of low, grassy hills. Then he started to move again. Chino Hills, apparently, wasn't big enough. Twenty-two times over the next 19 months, M6 made the journey back and forth under Highway 91. He became a street-smart lion, but the passage was always perilous. To reach the canyon, M6 had to work his way across two shrubless golf courses, which offered little in the way of protective cover, and past a stable. Before he could get to the freeway, he had to cross a double set of busy railroad tracks. This was, by any sentient being's measure, difficult terrain. Arc lights glared; traffic roared. . . . For M6 it had become a corridor of life and death. The lion could easily have been hit by a car (at least six mountain lions were killed by cars in southern California last year [2001]), kicked by a horse, or flattened by Amtrak. . . .

And life for this lion was about to get even harder. A developer had plans for 652 acres just south of the freeway: 1500 houses, plus all the usual gas stations and fast-food outlets that attend the birth of a neighborhood. Building and paving would sever the already tenuous connection to Coal Canyon. . . .

[Luckily for M6] Claire Schlotterbeck rallied support for mountain lions and other species that needed to pass through the bottleneck of Coal Canyon. After a two-year effort, her advocacy group, Hills for Everyone, persuaded the state to put up $14.7 million to help buy out the real estate developer who had planned on building condos south of Highway 91 and to purchase an additional 32 acres of land just to the north. "The department of transportation is restoring the underpass," says Geary Hund, a state parks department wildlife ecologist who worked with Schlotterbeck. "We'll get rid of the lights and put up some sound barriers. We'll pull up the pavement under the freeway and set up some fencing to steer animals off the highway and down under." Native vegetation will be planted on the compacted slope; Coal Creek's concrete waterway will be ripped up and half its natural flow restored. Instead of cement, he says, imagine willows and mulefat shrubs. Then imagine birdsong in the air, and butterflies.

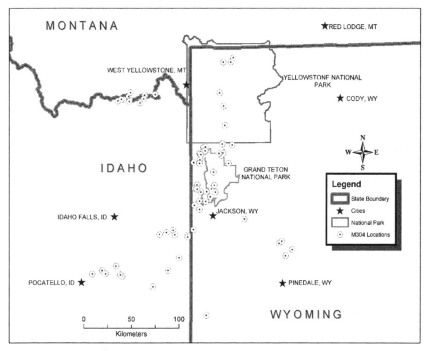

Figure 4.4. Recorded locations of a dispersing wolverine (*Gulo luscus*), M304, in the Greater Yellowstone Ecosystem over a period of about a year, taken from radio telemetry (vhf) and Global Positioning Unit (GPS) collars. During forty-two days of tracking by the GPS collar, the animal moved 874 kilometers (542 miles), and VHF locations indicate that it continued to move long distances after the GPS collar fell off (Inman et al. 2004; permission to use by Kris Inman).

Teton National Park and the Red Desert in Wyoming, a route used for over six thousand years. This route has several natural topographic bottlenecks (Figure 4.5; Berger 2004). Human developments such as roads, rural residences, and energy development could accidentally and permanently cut off this migration route. Researchers are working to characterize the exact migration route and the resources that the antelope need along the way. These efforts should help direct proactive planning that would permit that migration to continue into the future and ensure that this species does not disappear from the park.

Beyond the local or regional level, efforts are being made to identify and maintain connectivity on a much coarser level, such as across continents. Connectivity on this scale focuses less on individual species survival and more on keeping healthy ecological communities and their landscapes connected. Generally speaking, such corridors offer relatively protected habitat, increasing the total amount of habitat available for wildlife. In addition to providing habitat for wildlife and connectivity to core habitat areas, these webs of cross-regional corridors may allow unidirectional range shifts in the event of massive global change, such as climatic warming, thereby increasing the chances of long-term species persistence (Hobbs and Hopkins 1991, Harris et al. 1996, Burnett 1998).

Examples of such connectivity plans are the Paseo Pantera Project in Central America and the Yellowstone to Yukon Conservation Initiative (Y2Y), begun by the Wildlands Project, which addresses connectivity issues across North America. When introduced in 1990, the Paseo Pantera Project (www.afn.org/~wcsfl/pp.htm) represented a new approach to conservation in Central America with the vision of conserving a Central American Biotic Corridor more than 900 kilometers (558 miles) long. The Wildlife Conservation Society and the Caribbean Conservation Corporation initially spearheaded the effort, with seven countries agreeing to cooperate as signatories. Now many other groups have joined the effort, also referred to as the Mesoamerican Biological Corridor. Researchers modeled relative cost surfaces, or the optimal biological connections, to assess the best potential configurations of this corridor of reserves. The strategy was to buy land that would result in a series of linked reserves. Then ecotourism would help pay to protect those areas, and conservation groups would promote their conservation. One challenge has been maintaining collaborative efforts in multiple countries. Another difficulty is planning and managing ecotourism, especially in countries with little infrastructure to support tourism and few resources to monitor the impact of the indus-

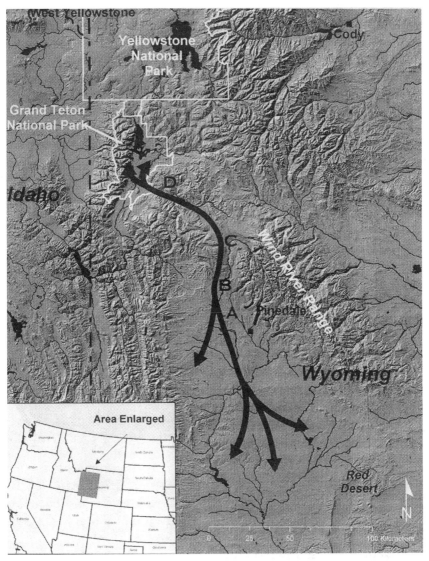

Figure 4.5. The migration route of pronghorn antelope (*Antilocapra americana*) of up to 270 kilometers (167 miles) between Grand Teton National Park and the Red Desert in Wyoming. Natural topographic bottlenecks along the route are noted: (A) Trapper's Point, a 0.8-kilometer (0.5 mile) natural constriction that has been used by antelope for 5,800–6,800 years; (B) Sagebrush Gap, a corridor 100–400 meters (328–1,312 feet) wide between a river floodplain and forest; (C) a high elevation divide; and (D) a 100–200-meter constriction between sandstone cliffs, a road, and the Gros Ventre River (Berger 2004; permission to use by Joel Berger).

try on biota (Trade and Environment 1996). Further, some politicians have tried to turn the notion into a catchall for rural development projects, and only a few countries had begun setting aside land or changing land uses two decades after the idea was born (Kaiser 2001).

Similar to Paseo Pantera, the Y2Y vision is intended to ensure that "the world-renowned wilderness, wildlife, native plants, and natural processes of the Yellowstone to Yukon region continue to function as an interconnected web of life, capable of supporting all of the natural and human communities that reside within it, for now and for future generations" (www.y2y.net). The Y2Y endeavor began with meetings convened in 1993 by the Wildlands Project and now includes nearly two hundred conservation groups as network members. It has identified seventeen critical core and corridor areas that need protection in order to achieve the network's collective vision (Figure 4.6). Some of the major challenges that Y2Y faces, similar to those of other large-scale efforts, are ensuring that network members work synergistically toward the same goals, that conservation actions are founded in sound science, and that science is directed toward regional considerations. An early criticism of Y2Y was that investments in science and modeling were not well linked with actual on-the-ground conservation action. Another issue that has been a challenge is extending the project vision across such a large spatial scale. A recent effort to identify critical areas and bring together member groups to focus on smaller sections of the larger Y2Y region addresses that issue. The project has been criticized for its focus on the Rocky Mountains, a less biologically rich region than other endangered ecosystems that contain a greater diversity of species. Much of the area is high-elevation rock and ice, reminiscent of many protected areas across North America. However, achieving the Y2Y vision will require successful conservation action in many of the lower-elevation valleys throughout the mountainous area. Despite criticisms of Paseo Pantera and Y2Y, these grand efforts stand to have a conservation impact on huge geographical scales.

In addition to launching Y2Y, the Wildlands Project spearheaded the scientific and design work on the Maine Wildlands Network, the Sierra Madre Occidental Biological Corridor (Mexico), and the Sky Islands Wildland Network in the U.S. Southwest—all ambitious conservation efforts on a large spatial scale. In all of these efforts, the vision is to identify, retain, and restore wildland network designs in various regions across North America using science-based tools to select critical sites (Noss 2003). While the Wildlands Project's goal is continental connectivity, its

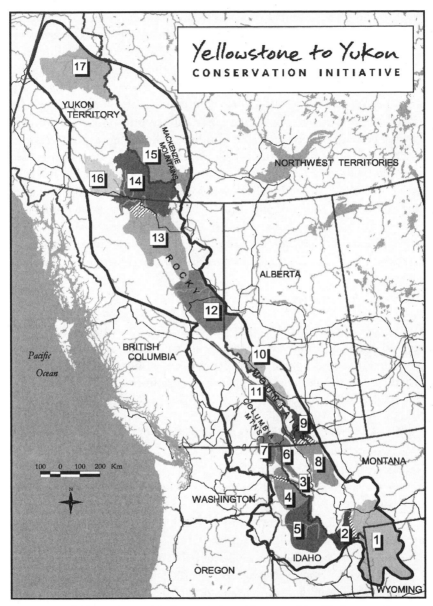

Figure 4.6. The seventeen critical core and corridor areas identified as needing protection in order to maintain connectivity along the spine of the Rocky Mountains, according to the Yellowstone to Yukon Conservation Initiative in North America. (Permission to use by the Yellowstone to Yukon Conservation Initiative.)

research and conservation activities focus on more local scales. Its approach includes objective methods for identifying critical core and potential corridor areas, educational activities, and working with local communities to find economic benefits for the conservation of those areas (Foreman 1999).

In the scenarios we have provided, corridors may facilitate daily or seasonal movement, dispersal, habitat connectivity, and long-term species persistence. It is easier to document daily and seasonal movement or habitat connectivity than enhancement of dispersal or long-term species persistence. We hope future studies will better investigate what factors may affect the utility of corridors to fulfill all of these goals. Some key topics in which research is needed include demographic changes before and after corridors are implemented and studies of dispersal patterns on fragmentation-sensitive species (Beier and Noss 1998). We have few realistic tools to mitigate habitat loss and fragmentation, so it is important that any projects document and monitor their efforts in order to inform future projects.

Biological Benefits

Ecological connectivity may benefit biota in a number of ways. Maintaining and restoring connectivity often means maintenance or enhancement of natural habitat, so one obvious benefit can be increased habitat. Additional habitat should permit greater species richness, as predicted by the theory of island biogeography (see discussion in Chapter 3; Noss 1987, Rosenberg et al. 1997, Gilbert et al. 1998). In support of that theory, a study in French Guiana revealed increased species richness of bat communities within forest patches connected by forested corridors compared to isolated patches (Brosset et al. 1996). Additional habitat also means the potential for more individuals within a species by providing more home range sites (Hess 1994). In Alberta, Canada, a study of forest birds revealed that forest corridors contained many resident adult birds and a large number of juveniles, suggesting that the corridors serve both as habitat and as conduits for dispersal following logging (Machtans et al. 1996). Increased numbers of individuals within a species can be very important for conservation of small populations that are constrained by human activities. In some cases, corridors may support actively reproducing populations that can disperse to other populations (Perault and Lomolino 2000).

Beyond the benefits of additional habitat, corridors can increase overall species' persistence compared to equivalent patches of habitat that lack connectivity. They do this by assisting in the movement of species among otherwise separate populations. By facilitating movements of species, corridors may serve to buffer groups of small populations from extinction by increasing persistence of species within a given patch and enabling recolonization of a patch after local extinction (Chapter 3; W. F. Laurance 1991, Beier and Loe 1992). For example, a study in Queensland, Australia, found higher survival of tropical rain forest mammals in forest fragments connected by corridors, indicating that corridors enhanced persistence (W. F. Laurance 1995). Similarly, in South Carolina, Bachman's sparrow (*Aimophila aestivalis*) was less likely to colonize isolated patches of habitat than patches that had a higher degree of connectivity (Dunning et al. 1995).

Increased connectivity can facilitate dispersal and thus increase genetic interchange among both plant and animal populations, reducing the risks of inbreeding depression (Beier and Loe 1992, A. F. Bennett 1999). Dispersal may increase levels of genetic variability within populations and reduce fixed differences between populations (Chapter 3). Even a low level of gene flow will avoid the chance fixation of deleterious genetic traits (Hedrick 1996). Genetic variability can increase species resilience to environmental change (Rosenberg et al. 1997), although some hypothetical concerns about dilution of locally adapted genes have been posed (Chapter 6). Spatial analyses of the impacts of fragmentation on a marsupial carnivore, *Antechinus agilis*, in Australia examined gene flow in continuous habitats, fragmented habitats, and fragmented habitats with corridors (S. C. Banks et al. 2005). The results offered evidence that the surrounding matrix was a barrier and that corridors provided increased gene flow among connected fragments.

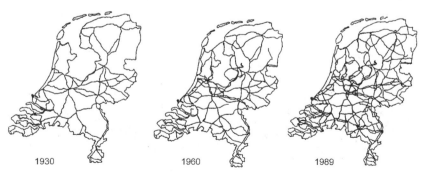

Figure 4.7. Changes in the road network from 1930 to 1960 and from 1960 to 1989 in The Netherlands (H&N&S landschapsarchitecten 1996; van Bohemen and Teodorascu 1997; permission to use by Heim van Bohemen).

In some cases, a road may bisect a wildlife population, limiting genetic interchange and contributing to wildlife mortalities. Figure 4.7 illustrates the density of roads in the Netherlands that potentially constrain movement of many species. Research shows that roads can isolate populations, and one example of this is near the Canada-U.S. border. Genetic techniques illustrated that the Selkirk Mountain grizzly bear (*Ursus arctos*) population in British Columbia is isolated from other populations by a road and associated human activity (Urquhart 2004). Even if species are willing to try to cross roads, they may end up as road kill. For example, within one month, road kills tallied in five U.S. states included 15,000 reptiles and amphibians, 48,000 mammals, and 77,000 birds; and those are likely underestimates (Havlick 2004).

Improving connectivity can maintain genetic interchange across roads, decrease wildlife deaths, and diminish vehicular damage. To those ends, crossing structures often are designed to serve multiple species. Some of

Figure 4.8. A culvert undercrossing included in the construction of twenty-two wildlife underpasses and two wildlife overpasses by 1998 because of concerns about the Trans-Canada Highway bisecting Banff National Park. (Photo by Anthony Clevenger.)

the most well-studied crossing structures are along the Trans-Canada Highway in Banff National Park. Concerns about the highway dividing the park led to the construction of twenty-two wildlife underpasses and two wildlife overpasses by 1998 (Figure 4.8). More crossing structures are being put in place as highway expansion and mitigation continue. The structures focus on species such as grizzly bear, black bear, wolf (*Canis lupus*), lynx (*Lynx canadensis*), elk (*Cervus elaphus*), and moose (*Alces alces*) (Clevenger and Waltho 2005). Research examining use of the Banff crossing structures indicates that those species have begun using them and that different species use different structures. Given increasing evidence that crossing structures are used by wildlife, many other countries have begun to consider the provision of wildlife crossing structures in road construction and reconstruction projects. The Netherlands has developed *ecoducts*, earthen passageways that facilitate wildlife movement over roads, which are successfully used by red deer (elk), as well as other species (Friedman 1997). The Montana Department of Transportation is putting in over forty fish and wildlife crossing structures along a 90-kilometer (56-mile) stretch of highway on the Flathead Indian Reservation, a project we discuss in more detail in Chapter 9 (Huijser 2004).

Corridors may also help dispersers avoid predation or human-caused death in attempting to cross matrix (Noss 1987). Because so few studies document dispersal of wide-ranging species, evidence of this is indirect. A study showing poor dispersal success of mountain lions across a heavily humanized landscape in Southern California indicates that human-caused deaths may limit potential dispersal (Beier 1995). Of nine dispersers, three were killed due to vehicle collisions, one was killed in an urban area by a police officer, and three died from disease and other natural causes. Whether corridors would decrease human-caused deaths in that case remains speculative, but there are few other management alternatives to consider. There is some evidence that corridors could direct the movement of species, the implication being that corridors could decrease human-related deaths and wildlife-human conflicts. For example, a study examining butterfly movements in South Carolina found that corridors between larger patches of habitat helped direct movements of specialist species among patches (Haddad 1999). This behavioral study demonstrated that focal butterfly species deflected off of corridor edges and were more likely to move between patches through corridors than across matrix habitat.

Continental corridors may be critical to species survival as climate fluctuates through the millennia. Evidence suggests that flora and fauna alike have shifted historically—for example, between glacial and interglacial periods (DeChaine and Martin 2004). In some cases, corridors may diminish the risks of extinction by facilitating range shifts among biota in response to catastrophic events or long-term environmental change (A. F. Bennett 1990, Beier 1995, Harris et al. 1996). Such relocation will be more challenging for species in the near future, with human development decreasing overall natural habitat and connectivity. A number of studies have attempted to predict how global climate change will affect species survival. They document how precipitation and temperature currently influence where species can occur and model where populations may occur in the future given potential changes. Models can show which populations would be most immediately affected by the changes. For example, a study of bighorn sheep (*Ovis canadensis*) that live on isolated mountain ranges in Southern California indicated that climate change could determine which populations persist in the long term (Epps et al. 2004). While it is clear that habitat fragmentation increases the risk of extinction in the case of global warming, the role corridors might play for species adapting to global change remains untested and may not be significant if change is rapid and existing communities do not have time to adapt. Although few data exist to support the prediction that cross-regional corridors would assist long-distance movement in the event of climate change (Hobbs and Hopkins 1991), it is hard to imagine any realistic alternative that would be conducive to species persistence.

Finally, corridors can help retain healthy functioning ecosystems. For example, where riparian zones are buffered to create natural corridors, the broad strip of natural habitat can retain the overall functioning of river systems. Corridors also can keep predators in habitat fragments, the loss of which can result in a cascade of ecosystem impacts (Power et al. 1996). Further, they can help maintain species and essential services such as pollination (e.g., Kremen and Ricketts 2000), and corridors everywhere can be seed sources for revegetation and recruitment of the diversity of plant species.

Benefits to Humans

Designing, maintaining, and restoring a network of connectivity across a landscape can directly benefit humans, as well as biodiversity. If such zones of connectivity are open to public access, open spaces can be important places for recreational hiking, biking, and relaxing. For example, the city of Boulder, Colorado, offers recreational and outdoor opportunities in the surrounding open spaces and mountain parks (www.ci.boulder.co.us/openspace). Also, its urban greenway trails serve as an alternate commuting route for those not driving cars. Countryside corridors offer human amenities as well. Naturally vegetated field margins not only increase the aesthetic appeal of the countryside, they can also incorporate horseback riding, hiking, and biking trails for recreation (Fry 1994).

Depending on the species, suite of species, or natural community that a corridor is supposed to support, combining connectivity for biodiversity with human recreation may or may not be appropriate. Research examining the influence of recreation on wildlife indicates that many species of wildlife are sensitive to the presence of humans and may not use an area heavily used by humans (e.g., A. R. Taylor and Knight 2003). Also, foot, horse, and bike travel are modes of weed transport. Weeds brought in by human activities can out-compete native plants, including food sources for wildlife, compromising the integrity of the corridor (discussed further in Chapter 6).

Beyond recreation, rural and urban greenways play an important role in limiting urban expansion and sprawl and retaining distinct boundaries around different urban areas (Ahern 1995, Kubeš 1996). For example, one of the objectives of Sonoma County's Agricultural Preservation and Open Space District (SCAPOSD; California) is to identify and retain community separators and keep various communities from growing together (SCAPOSD 2000). Such greenways may also offer aesthetic enjoyment and recreation. The scenery can be an important part of the attraction of people to a community or region and can increase property values (Noss 1987). The effectiveness of greenways for containing development within urbanized areas remains largely untested, as development may just leapfrog beyond greenbelts. Whether greenways are effective at containing development likely relates to the strength of other tools that can direct or limit growth, including zoning and other regulations, as well as private land conservation tools (see Chapter 8).

Corridors also provide free ecosystem services. Retaining buffer corridors on steep hillsides, for example, can limit hillside slumping, landslides, and erosion. Greenbelt corridors can also limit pollution, such as from busy highways into adjacent neighborhoods (Noss 1987, Kubeš 1996). Likewise, retaining buffers along creeks and around wetlands can help sustain the natural water-filtering process. Because human disturbances change the flow of materials, riparian corridors can serve to moderate flows of such materials as sediment, fertilizer, toxic residue, and pesticide into river systems (Fry 1994, Ahern 1995). Where anthropogenic inputs to streams are regulated because of the presence of endangered species, or to protect human water supplies, natural vegetation filters are far less costly than installing high-technology water filtration systems that are often used to ensure that water quality standards are met.

Corridors or strips of natural habitat also can be beneficial in agricultural systems. Hedgerows and other linear habitats can help limit soil loss due to wind and water erosion. In addition, corridors can help retain snowpack in windy areas, increasing total water accumulation and storage. Especially noteworthy is a long-term research program extending more than thirty years on agricultural landscapes in western Poland (Wielkopolka) led by Lech Ryszkowski. This program found that belts of vegetation such as shelterbelts, strips of meadow, and hedgerows in agroecosystems provided multiple benefits. Rainfall increased in crop areas adjacent to shelterbelts, and winds were ameliorated. Lateral movements of pesticides were reduced; soil and air temperatures were cooler in crops during hot weather, reducing evapotranspiration. Adjacent belts of vegetation helped reduce soil erosion by water and wind, helped economize overall water needs, and decreased leaching of nutrients such as nitrate ions, which reduced the need for added fertilizers (Ryszkowski et al. 1996). The reduced export of pesticides and nutrients had the secondary effects of reducing pollution in groundwater and adjacent watercourses and preventing eutrophication of nearby lakes and rivers. In addition, landscapes with these corridors had much higher densities and diversity of animals, including game animals, pollinators, and predators of crop insect pests. Shelterbelts offered a supply of wood, shade, and aesthetic amenities; hedgerows served as effective fences for livestock (Ryszkowski et al. 1996). Comprehensive investigations such as this are leading the way toward management of agroecosystems for sustainability, maximum production, control of nonpoint sources of pollution, and conservation of native biota.

Corridors within agricultural areas can provide other direct services to farmers. Where agricultural lands and residences are adjacent to one another, natural vegetation buffers at field margins can reduce the drift of pesticides into residential communities, into waterways, and between fields (Fry 1994, Ryszkowski et al. 1996). Because monoculture farming is most susceptible to pest damage, retention of natural vegetation along field margins can reduce the effect of pest populations in two ways. First, it can intercept searching behavior of pests, especially those with poor dispersal ability, and thereby reduce pest damage (Fry 1994). Second, field margins can serve as reservoirs of natural enemies of crop pests (Forman 1995), potentially reducing the need for pesticides. Many predators that are helpful in reducing pests in agricultural areas have higher densities close to the edge of fields (Fry 1994, Nicholls et al. 2001). One study, for example, showed that the presence of natural vegetation corridors suppressed populations of leaf- and stem-sucking pests in soybean mono-cultures, although those corridors did not diminish defoliator pests (Rodenhouse et al. 1992).

Species harbored within more natural vegetated corridors may also play important pollinating roles. A study in California demonstrated that maintaining functional migratory corridors for pollinators that are nega-tively affected by habitat loss and fragmentation helped to sustain native pollinator ecosystem services in agricultural and natural systems (Kremen and Ricketts 2000). Similarly, a study in Costa Rica found that coffee plantations within 1 kilometer (.62 mile) of a forest had 20 percent higher coffee yields because of improved pollination (Ricketts et al. 2004). Fur-ther, the results of a large-scale experiment in Florida demonstrated that thin strips of habitat among patches facilitate two types of plant-animal interactions, namely pollination and seed dispersal (Tewksbury et al. 2002).

In summary, corridors can take many forms and can be beneficial for natural and human systems on a variety of different scales. From road-crossing structures to watershed corridors to continent-wide connections, they can facilitate dispersal and migration and increase the overall quan-tity of habitat available, helping individuals move and populations retain connectedness.

CHAPTER 5

Role of the Matrix

In this chapter, we will expand our perspective to explicitly consider the context in which we find the metapopulations and metacommunities described in Chapter 3. That context is the *matrix*, the various kinds of communities in all of their physical and biotic dimensions that surround habitat patches and can influence the metapopulations and metacommunities of concern. In anthropogenically modified landscapes, the matrix is often the human-influenced or human-produced communities that cause the fragmentation of native communities and habitats in the first place (Figure 2.1). But even naturally heterogeneous landscapes contain a mosaic of patches of the focal community type intermixed with other community types. Note, however, that the designation of a community as matrix must be in relation to the focal or target community of interest. If we shift our attention to a different community type, what was previously the metacommunity of interest may become part of the matrix for the new subject of our attention.

Basically, all of the communities outside of the community type of special interest collectively constitute the matrix. As it often comprises many different kinds of communities, the matrix can be complex, indeed. The patches in the mosaic will be explicitly arranged across space in the landscape system being investigated, and each will have a specific shape and spatial dimensions. Such an array of communities will be influenced by physical features such as topography; substrate types; and exposure to wind, waves, water currents, and various other weather factors. Some of the matrix or all of it may be human-modified communities such as agricultural fields of various kinds, clear-cut forests, grazed grasslands, and towns. Various barriers to movements of organisms may be present, as, for example, roads, fences, irrigation ditches, airports, and urban developments. Finally, the matrix can be dynamic, with seasonal changes of many kinds, as well as changes caused by successional processes, fire regimes, and human activities. As we will see, organisms living within the community fragments of interest will be profoundly influenced by the size and qualities of the matrix to which they are exposed. For example, Drapeau et al. (2000), in a study of bird communities in northwestern Quebec, illustrate this point. They demonstrated that the nature of avian assemblages was about equally influenced by habitat features and by the landscape context in which they occurred. Similarly, Reunanen et al. (2000) conclude that for flying squirrels (*Pteromys volans*) to persist in the boreal forests of Finland, it is not sufficient to provide large patches of mature mixed deciduous and conifer forests (optimal habitat); the patches must also be connected by dispersal corridors of young secondary forest.

The matrix poses special challenges for those wishing to mathematically model the behavior of organisms in complex spatial arrays such as metapopulations or metacommunities. While accepting that actual matrices are likely to be quite complex, modelers will necessarily have to simplify them in order to limit the number of parameters incorporated in particular models. We have already seen how the original metapopulation models by Levins (Chapter 3) assumed a dimensionless and uniform matrix that could be characterized by a single parameter, namely, the probability of an emigrant being successful in founding a new colony in an empty patch. Other models take distances between patches into account but assume that only one uniform community type is involved. Still others may assume that matrix influences extend only a limited distance from fragment edges. Such simplifications, while necessary on practical grounds, will influence the model outcomes to varying degrees. An unusual modeling study was performed by Cantrell et al. (1998), in which they treated the matrix not as a constant but as declining in quality. Their model considered the fate of two competing species living in habitat patches and found that the competitive coefficients between the species shifted around as the matrix deteriorated and could even reverse in sign, so that the previously subordinate species became competitively dominant. This dynamic resulted from differing negative influences on dispersers of the two species in the matrix as it degraded.

Research on the role of matrix is grounded in the subdiscipline of landscape ecology, and that is where we will begin our discussion. The rest of the chapter will examine the role of matrix in influencing connectivity among fragments or patches, the importance of edge effects generated by specific juxtapositions of community types, effects of matrix on the population dynamics of target species, and the influence of matrix on the invasions of exotic organisms. Although these are all important aspects of conservation, they are often ignored or insufficiently appreciated. For example, Calabrese and Fagan (2004) provide a useful discussion of various connectivity metrics but make no explicit mention of the critical role of matrix in understanding landscape connectivity. Lindenmayer and Franklin (2002), on the other hand, give us a thorough treatment of the importance of matrix to the conservation of biodiversity in forests.

Landscape Ecology

Considering the role of matrix brings us explicitly into the field of landscape ecology, the dimensions of which are commonly perceived only hazily. It is important therefore to provide a brief overview of the nature of this important subdiscipline. This is the branch of ecology that is concerned with the interactions among patches of different community types. As such, the spatial scale of systems studied by landscape ecologists is generally quite large compared to studies of single communities or populations. The spatial dimensions, however, will often be no larger than those of metapopulations or metacommunities. In fact, a landscape of interest need not necessarily be very large at all. Small species might inhabit quite small patches, and the relevant landscape could be composed, for example, of an array of downed logs in a forest, a cluster of cow droppings in a pasture, or even a small population of host individuals for a parasite.

The ecological concept of landscape grew from several traditions (Lidicker 1995b). One of these dates to agricultural reform movements in the late nineteenth century, with central European countries such as Poland and what is now the Czech Republic centrally involved. Ideas such as windbreaks, crop rotations to maintain soil fertility, and methods to prevent soil erosion were examples of early landscape-related ideas. As the discipline of ecology developed through the twentieth century, this tradition expanded and generalized, adding the new mission of incorporating human-modified systems into mainstream ecology. This was a difficult task, as most ecologists were interested in studying intact "natural" systems and explicitly excluded human-modified ecosystems from their purview. They thus drew a firm line between basic and applied ecological research. Only in recent years, with the universal recognition that it has become increasingly difficult to find natural communities to study, has the barrier largely broken down. Of course, a second factor in this has been the realization by ecologists that they must get involved in studying human influences on our planet's landscapes or there will likely be no future for human civilization, much less for ecological research.

From this agroecology perspective, a scientific definition of *landscape* was formulated. According to this tradition, landscapes were relatively large portions of the living world (the biosphere), often expressed as areas of one or more kilometers in linear dimensions. Such large areas would generally contain a complex mixture of community types, and in most versions of the definition it was necessary to include anthropogenic features as well. However, large spatial expanses do not necessarily guarantee that community heterogeneity will be included. There are still large portions of the oceans, large lakes, tundra, and boreal forests where community homogeneity prevails at the 1-kilometer scale. For that matter, there are also urban expanses that far exceed 1 kilometer in diameter and generally are homogeneous with respect to community composition.

A second intellectual root of landscape ecology began in the mid-1980s, when the modern era of the discipline really took hold and rapidly spread around the world (Hansson 1995, Lidicker 1995b). This was a time when there was still a major disconnect between mainstream ecology and this new interest in "landscapes," especially regarding those landscapes that contained humans and their artifacts. Some recognized that these emerging landscape ideas were too important both intellectually and for the future of humanity to remain distinct from the basic ecological research tradition. A new definition for landscape was needed to bridge this divide (Lidicker 1988b, 1994, 1995b). It seemed that what was special about landscapes was not so much their spatial dimensions or the presence of people, but the inclusion of two or more community types within the ecological system under investigation. This approach does not constrain the size of a landscape or necessitate the inclusion of human-dominated or human-modified communities or artifacts. The argument was that viewing landscapes simply as possessing two or more types of communities would make it possible to connect landscape ecology intellectually to the rest of ecology.

Ecologists were already fairly comfortable with the holistic notion that individuals, populations, and communities represented three levels of increasing biological complexity. Each new level of complexity incorporated the levels below it as the parts making up the new system. Moreover, each level possessed properties that were not features of the level below it. That is, populations, for example, were characterized by variables such as age structure, sex ratio, mortality rates, and so on. These are clearly not features of the individual organisms that made up the population but rather are emergent properties—a fundamental ingredient in holistic philosophy. Landscapes also possess emergent properties, in part as a consequence of different communities living next to each other. Some of these properties are listed in Box 5.1.

Basically, landscapes are composed of different kinds of communities, and these may or may not include human-modified and human-produced communities along with human artifacts such as roads, dams, canals, and settlements. Aside from the ecological processes going on within communities, landscape processes involve fluxes among the interconnected communities (Box 5.1). As far as organisms are concerned, this means that they must cross edges between community types, move across matrix, and

Box 5.1.

Emergent Properties That Characterize Ecological Systems of Two or More Community Types

- Types of communities present
- Diversity of community types (numbers of patches and proportions)
- Spatial extent of various community types (and proportions)
- Biomass (by community type and overall)
- Spatial configuration of patches (dispersion, sizes, shapes, juxtapositions)
- Ecotonal features (edge effects)
- Connectedness (links among patches, corridors, barriers)
- Interpatch fluxes (energy, nutrients, organisms, information)
- Stability (resilience, constancy, predictability)
- Long-term trends (succession, degradation, exotic invasion, extinctions)
- Energetic properties of individual community types and the entire array of communities (productivities, decomposition rates, turnover times, trophic efficiencies)
- Historical context
- Anthropogenic influences

travel through corridors. These are topics we will explore in this chapter and in Chapter 6. For more thorough treatments of landscape ecology, see Forman and Godron (1986), Lidicker (1995a), Sanderson and Harris (1999), and Turner (1989).

Traveling the Matrix

If landscape function concerns connectedness among the communities that make up the landscape, movements of organisms among the habitat patches, that is across the matrix, are of fundamental importance. These movements can be dispersal, within-home-range movements, or exploratory excursions. Longer-range movements will primarily be dispersal (Chapter 3), and those are the movements that will be of most concern to conservation planners. To begin with, it is critical to acknowledge that real matrices can be very complex, composed of a variety of different types of communities, variously arranged spatially, and of differing dimensions. Avenues of relatively easy travel through a matrix, that is corridors, may or may not be present and available to those with wanderlust. Moreover, each species in a community will perceive the matrix in its own way and respond uniquely. The assessment of matrix permeability to movements will therefore not generally be easy to discern and will be especially difficult to predict in a management context.

It may be useful to think of matrix surrounding a particular community patch as being on a gradient with respect to its traversability for a given species living in the patch. We can devise six levels or grades of permeability:

- No impediment exists except for distance to the next available patch.
- No impediment exists at certain times; various levels of impermeability exist at other times. Favorable times might be seasonally organized, multiannual events, or connected to major unpredictable perturbations. For example, a rare flooding event may connect the headwaters of one stream system to another.
- Minor inhibitions to travelers exist, magnifying distance effects. These, of course, can also be intermittent.
- Moderate impediments to movements exist.
- Strong barriers to movements exist, but some movements do occur. Successful crossings may be limited to certain age, sex, or health groups or to seasons.
- Complete barrier to movements exists.

A particularly nefarious complication with this gradient would occur when a matrix provides moderate or good permeability but is bisected by a non-permeable barrier such as a major highway or river that prevents travelers from reaching otherwise available patches. S. G. Laurance et al. (2004) have shown that even narrow, low-traffic roads in Amazonia can greatly inhibit movements of many rain forest species of birds. This is similar to results reported by Diamond (1972, 1973) for rain forest birds in New Guinea and southwest Pacific islands

Permeability or traversability of matrix habitat will depend on a species' access to the matrix (more about this later), the quality of the matrix with respect to survival and facility of movements, and the distance to neighboring patches. The variables of distance and quality interact, as shown in Figure 5.1. If distances are short and quality is good, individuals may even be able to combine several patches into a single home range (Figure 5.2; Andreassen et al. 1998, Dunning et al. 1992, Kozakiewicz and Szacki 1995, Noss and Csuti 1997). Therefore, a potential traveler will need to assess both the qualities of the matrix and its dimensions. A poor-quality matrix may be crossed if the distance to the next patch is not too great. Longer travels may be ventured if the matrix is more favorable. Several studies with voles (*Microtus*) in experimental landscapes illustrate this dis-

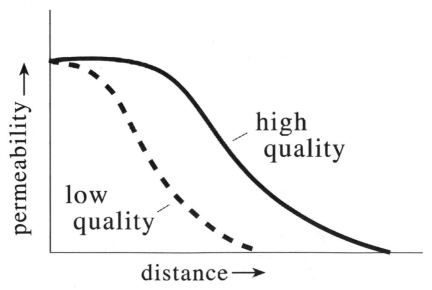

Figure 5.1. Permeability of matrix as experienced by dispersing organisms as a function of distance to neighboring habitat and matrix quality.

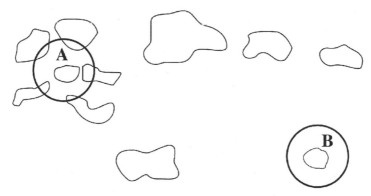

Figure 5.2. Several patches of habitat assembled into a single home range (A) and a home range composed of a single isolated patch (B).

crimination. Distances of 1 meter (39.37 inches) of matrix are readily crossed, whereas 4 meters is considered risky but still crossed frequently and 9 meters is rarely crossed (Andreassen, Ims, and Steinset 1996, Lidicker and Peterson 1999, Wolff et al. 1996, 1997). An important variable here is the ability of an organism to perceive distance across the matrix.

Of paramount importance is the recognition that matrix permeability is particular to the various species in the community. What is easily crossed by some may be impenetrable to others. Thus, matrix may serve as a filter, allowing good connectivity for some components of the focal community but not for the rest. When this situation applies, and it is likely to be common, progressive local extinctions of disconnected species will allow community structure and function to drift away from their initial conditions. Eventually, extinctions occur, and the original community type will transform into something else. These delayed extinctions can be a slow process and hence difficult for managers to detect. Community patches in some landscape contexts can therefore appear healthy but actually be moribund. This condition has been called the *extinction debt*. Ultimately, it must be paid.

Differential use of the matrix is illustrated by Gascon et al. (1999), who studied birds, small mammals, frogs, and ants in forest fragments in central Amazonia. They found that populations of species using the matrix were stable or even increased following fragmentation, but those that did not use the matrix declined or became extinct. Similarly, Ricketts et al. (2001) documented the moth biota in a 227-hectare (561-acre) forest fragment surrounded by four types of agricultural development in Costa

Rica. The various species of moths moved different distances into the agricultural matrix, and overall the moth species richness was greater within one kilometer (0.62 mile) of the forest than at more distant sites.

Because matrix may filter out those species less able to travel to other patches, it is appropriate to consider what features of organisms favor movement abilities. The ability of organisms to travel is termed their vagility. Generally, large, nonsedentary species are good travelers. It helps if they are habitat generalists and thus comfortable in a variety of community types. Species with superior movement capabilities include runners and fliers. Individuals who are trophic generalists or who can go for long periods without eating can relatively easily cross nonhabitats. Some small species, and especially seeds, can attach themselves to larger species and get transported by them (phoresy). An interesting case involving blind snakes (*Leptotyphlops dulcis*) illustrates how bizarre phoresy can be (Gehlbach and Baldridge 1987). These snakes are picked up by screech owls (*Otus asio*) and placed in their nests. Here they feed on nest arthropods and ectoparasites of the nestling owls. Young owls grow better and suffer less mortality when snakes are resident in their nests. When the owls fledge, the snakes leave the nest and discover that they have dispersed to a new place. We humans are, of course, world champion phoretic agents. We intentionally and unintentionally transport living creatures all over the world, even across major ocean basins. Sometimes these transfers benefit humanity, such as when the potato was brought from South America to Europe. Most, however, are detrimental to agriculture, forestry, human health, and native biotas, and this has become a huge conservation problem, second in importance only to loss of habitat for species (Pimentel 2002).

Even very small creatures may be good dispersers. Many micro-organisms and even very young spiders can use air or water currents to passively drift to new patches. Other things being equal, asexual species will have better odds of successfully establishing in an empty patch than those that require more than one sex or mating type to reproduce successfully. Those organisms with high reproductive rates will also improve their rate of success by sending out large numbers of dispersers. One of the largest litter sizes found among voles of the genus *Microtus* occurs in the taiga vole (*M. xanthognathus*) of northern Canada and Alaska. Its preferred habitat is recently burned spruce (*Picea*) forests. As succession proceeds following fires, the forest gradually becomes unsuitable for the voles. They succeed by having offspring that find new burned patches in unpredictable locations (Wolff and Lidicker 1980). Species involved in complex, especially obligate, social relationships are unlikely to be able to penetrate matrix successfully. And if they do successfully reach a new patch, they will likely be fended off as unwelcome immigrants and not be able to integrate into existing social groups. More subtle but potentially critical attributes are a species' propensity to enter nonhabitat and its capacity for finding and using movement corridors. We will discuss these two issues later.

Conservation planners can, with caution, as many exceptions are known (Lidicker and Koenig 1996), use morphological, behavioral, and life history features like those discussed to effectively design landscapes that will connect community patches most advantageously. Conversely, they can use them to predict where problems are likely to occur.

Edges and Edge Effects

As discussed in Chapter 2, human-induced fragmentation increases edge habitat across the landscape. Edges, of course, are not unique to communities. The organized nature of living material leads to boundaries or borders between organized units, and these are evident from the organelles that form the constituents of cells to all the more inclusive and complex levels in the hierarchy of life. Ecologists are concerned with these organizational borders, now called edges, at the levels of the individual organism, the population, the community, and on up to the biosphere. However, in this chapter, it is the edges between different kinds of communities that will engage our attention. Matrix is one side of this edge, and so its role needs to be examined.

We can assume that the organizational forces within a community type are commanding, and that they have a prevailing influence on generating the character of that community. When two different communities share a common boundary, fluxes across that boundary develop. These may be movements of abiotic materials, of organisms or their gametes, and even of information. When a rabbit peers out of its brushy hiding place and observes a predator walking by, information has been transferred across the border. One abiotic flux generally ignored is that of artificial light. Bird et al. (2004) demonstrate that artificial lighting adjacent to the habitat of beach mice (*Peromyscus polionotus*) in Florida significantly inhibited their foraging activities. Such intercommunity fluxes, while less abundant than those within each community, can nevertheless have important influences on neighboring patches.

Movements of nutrients across community boundaries can be extremely important. It has long been recognized that streams, rivers, and estuaries often receive large inputs of nutrients from adjacent terrestrial communities (e.g., Wallace et al. 1997). Less appreciated are the sometimes significant movements of nutrients out of the aquatic environment to the terrestrial. A particularly spectacular example is given by black bears (*Ursus americanus*), who remove salmon (*Onchorhynchus* spp.) from streams in the Pacific Northwest and carry them inland up to 150 meters (almost 500 feet). In one study (T. E. Reimchen, pers. comm. 1997), bears removed 63 percent of a salmon run and left half of that in the forest to benefit a myriad of scavengers, decomposers, and plants (Ben-David et al. 1998).

Thus, when patch meets matrix, new dynamics are generated, and these are called *edge effects*. The border area, which manifests these effects, is often called an *ecotone*. The idea of edge effects is often attributed to Aldo Leopold (1933), at least by animal ecologists (Lidicker 1999). Leopold considered edge effects to be beneficial features of landscapes. Edges were often sites of high productivity and enhanced biodiversity and so were much desired by wildlife managers. The notion of ecotone dates to the end of the nineteenth century (Clements 1897). Boundaries between plant formations (that is, communities) were thought of as "tension zones," often featuring enhanced productivities. Thus it was that ecotones became associated with new or emergent properties, not necessarily predictable from knowledge of the two adjacent communities separately. Another feature of ecotones is the possibility that their biodiversity will be enhanced by the occurrence of edge species, that is, organisms whose primary habitat is the ecotone itself.

More recently, and concomitant with increasing anthropogenic fragmentation of native communities, the realization has grown that edges can have negative effects on the participating communities as well as positive (W. F. Laurance 1997, Yahner 1988). Particularly vulnerable are species that live in the interior of patches and avoid or do poorly on edges, and there are increasing numbers of examples reported of this behavior. Since edge zones may or may not be used by species living in the patch, the area encompassed by edge may need to be subtracted from the total patch area to give the effective patch size. This results in reduced patch areas and therefore smaller population sizes for the inhabitants (Figure 5.3), increasing the risks of all the negative effects of small population sizes (Chapters 2, 3). If patch residents actually avoid going into the ecotone, it will re-

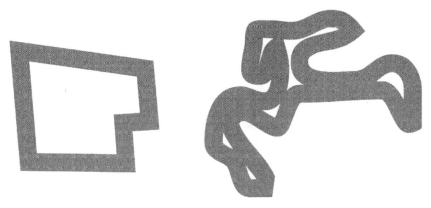

Figure 5.3. Two differently shaped habitat patches of the same size showing different edge encroachment. The edge zone is indicated by the gray borders. The patch on the right has a vastly greater edge-to-area ratio than the one on the left and therefore has a much smaller effective patch size for edge-intolerant species.

duce their chances of dispersing out of the patch. In fact, we can grade a species' behavioral response to edge ranging from refusal to even enter the ecotone, and hence being unable to find the patch edge and its adjacent matrix, all the way to entering the matrix without inhibitions (Figure 5.3 in Lidicker and Koenig 1996). And, of course, a given edge will provoke various responses in different species.

Whether good or bad, we can anticipate that a given edge will produce a multitude of effects on the various species that encounter it. One category of effects is the extent to which an interior species will use ecotones either for residency or travel, or perhaps not at all. If it does live in the ecotone, is its performance in terms of mortality and reproduction equivalent to that of individuals living interior to the edge? Driscoll and Donovan (2004) report an interesting complexity in which the reduced nesting success that characterizes wood thrushes (*Hylocichla mustelina*) nesting in forest edges bordering agricultural fields (in central New York) disappears if the fields are merely patches embedded in otherwise continuous forest.

Moreover, examples of reduced use of edge by nesting birds are increasingly being reported for a variety of habitats (see examples in Lidicker and Koenig 1996). The ovenbird (*Seiurus aurocapillus*), a deciduous forest interior species, nests successfully but at 40 percent lower densities within an edge zone that extends 150 meters into the forest (Ortega and Capen 1999), and King et al. (1996) report that nest survival was higher inside of a 200-meter (656-foot) edge zone. In tropical forests of Uganda, some interior species suffered reduced densities up to 450 meters (1,476 feet) from edges (Dale et al. 2000). In boreal forests, also, four species of birds did not respond to territorial calls across a forest gap, and in fact responses declined starting 40 meters (131 feet) from the gap (Rail et al. 1997). Edge effects are not limited to forest birds, as demonstrated by a particularly informative study on bobolinks (*Dolichonyx oryzivorus*), a grassland species. Nesting density and success rates were reduced within 100 meters (328 feet) of forest or hedgerow edges, and birds with failed nests on edges moved farther from the edge when renesting. Roads elicited decreased nesting densities but without reducing success rates. Finally, edges with old fields or pastures were not avoided, and nests on those edges enjoyed success as good as or better than interior grassland nests (Bollinger and Gavin 2004). Albrecht (2004) reported on edge effects in shrubby wet meadows bordering on crop fields in the Czech Republic. There scarlet rosefinches (*Carpodacus erythrinus*) exhibited 41 percent nest success within 100 to 200 meters of the edge compared to 83 percent survival for interior nests. Productivity of nestlings per capita was also 63 percent higher in interior nests.

A second kind of edge effect affects the ability of species to move through the edge. This effect is sometimes called edge permeability or edge hardness (Stamps et al. 1987a, b, Duelli et al. 1990). Permeability is a feature central to models that involve the relative amount of edge and its effects on dispersal dynamics (Buechner 1987, Stamps et al. 1987a, b). It is important, therefore, that that parameter be accurately measured. A recent report by Sieving et al. (2004) describes a fascinating situation in north-central Florida in which forest interior species of birds readily crossed an edge into open habitat when foraging in mixed-species flocks. Specifically, when the tufted titmouse (*Baeolophus bicolor*) was present, other species were more inclined to move into open habitat to mob a stuffed screech owl (*Otus asio*) equipped with recorded calls. The authors speculated that this facilitation resulted from a perception of reduced predator threat in the presence of the socially dominant and highly alert titmouse. This raises the interesting possibility that land managers in this region could actually increase the connectivity of a landscape for forest interior birds by encouraging the presence of tufted titmice.

The modeling of matrix permeability should logically incorporate edge permeability, matrix quality, and distance that must be traveled. This is not an easy assignment. If edge enhances dispersal, it is likely that that effect is caused by the edge itself. On the other hand, if crossing movements are inhibited, the effect may be a product of both edge characteristics and matrix characteristics. Relevant here for modelers as well would be knowledge of the ability of focal organisms to assess matrix quality and its dimensions. Poor judgment about the matrix that needs to be traversed will negatively affect the travelers' chances for success.

Both edge use and edge permeability can be expressed in a summarizing metric of performance measured across the edge (Figure 5.4). Performance can be measured in terms of behavior, physiology, or numbers, as seems appropriate. Two major categories of edge effects are immediately apparent. One of these comprises cases characterized by the absence of emergent properties for a given target species. In such cases, the response of organisms across the edge can be explained strictly by the organism's response pattern in the two adjacent communities separately (Figure 5.4A) and is referred to as a *matrix edge effect* (Lidicker and Peterson 1999, Lidicker 1999). Diagnostic of this pattern, the organism's behavior on the edge changes abruptly from that associated with one community to that found in the other community, or the change is more gradual but symmetrical across the edge, or the change is accurately reflective of the intermediate character of conditions in the edge. In the last case, the response pattern corresponds to the degree of mixing of the two communities. The second type of edge effect is associated with emergent properties; that is, the edge will elicit a response that is not predictable from knowledge of the organism's behavior in the two community types when they are not adjacent to each other (Figure 5.4B). This category of behavior is called the *ecotonal edge effect* (Lidicker 1999, Lidicker and Peterson 1999). The new response to edge may be an enhancement of function, a diminution of response, or an asymmetrical pattern not attributable directly to the mixing of two community types.

In practice it may not always be easy to distinguish these two categories of effects. For example, pattern b in Figure 5.4A may appear quite similar to patterns c and d in Figure 5.4B. Operationally, the null hypothesis will be the matrix effect. It can, in principle, be predicted from known performance of a species in the two habitats separately, plus the measured blending of the communities on the edge. Any significant deviation from this prediction will suggest an ecotonal effect and indicate directions for further investigation of the phenomenon.

Matrix effects would be anticipated when two adjacent communities are quite different, such as with aquatic-terrestrial borders or with abrupt changes in soil type. In a study of forest-farmland edges in Illinois, Heske

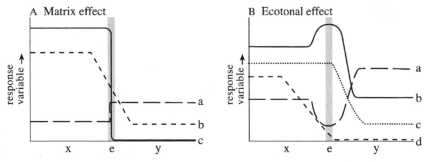

Figure 5.4. Idealized graphs representing matrix and ecotonal edge effects. Response variables (*y* axes) such as numbers and intensity of use are plotted against space traversing two habitat patches, x and y, and the edge between them. The matrix effect (A) is depicted without emergent properties, and the ecotonal effect (B) shows the presence of emergent properties (Lidicker 1995b).

(1995) found only matrix effects for five species of carnivores and four species of rodents. On the other hand, Mills (1996), investigating small mammal responses to edges in Douglas fir (*Pseudotsuga menziesii*) forest with clear-cuts in Oregon, found that among four species, three different edge responses occurred. One species (Townsend's chipmunk, *Tamias townsendii*) exhibited a matrix effect; two species (red-backed vole, *Clethrionomys californicus*, and deer mouse, *Peromyscus maniculatus*) showed ecotonal behavior, with effects extending 45 meters (148 feet) into the forest; and one (Trowbridge's shrew, *Sorex trowbridgei*) showed no recognition of an edge at all. In the absence of fire, taiga voles in Alaska persist mainly on the edges of taiga forest and swales of horsetail (*Equisetum*). This combination of communities supplies both good burrowing conditions (above the permafrost) and a supply of rhizomes essential for successful overwintering (Wolff and Lidicker 1980). Similarly, adult male cotton rats (*Sigmodon hispidus*) in South Carolina preferentially overwinter where patches of good cover (*Rubus* sp.) adjoin grasslands, which presumably have better food resources (Lidicker et al. 1992). Sex-biased use of matrix is also shown by the Eurasian flying squirrel (*Pteromys volans*) in Finland, where males readily crossed a matrix of low-quality forest unsuitable for breeding, whereas females foraged in the matrix but did not cross it (Selonen and Hanski 2003).

One aspect of edges that is important in landscape models and reserve design is the length of edge relative to area of fragment. Area can be relatively easily measured, but edge length is definitely tricky. The measured length of any boundary that is not absolutely straight will depend on the length of the measuring unit used. The more complex a boundary is, the shorter the measuring segment must be to capture that complexity. Fractal geometry is a method for analyzing the shape and hence the perimeter of complex objects, which is exactly what ecologists typically contend with. For complex edges, the measured length increases as the measuring segment decreases. The larger the measured perimeter relative to the size of the estimator segment, the more complex the boundary and the greater its fractal dimension. This relationship is illustrated for the country of Norway in Figure 5.5; the same principle applies to small habitat patches. For the biologist and land manager, what is important is that the measuring segment be relevant to the organisms and processes of interest. For example, if one is interested in landscape boundaries of relevance to a wide-ranging carnivore such as a gray wolf (*Canis lupus*), an edge measurer of tens of kilometers might be appropriate. On the other hand, for a beetle, one would have to use a length of a meter or less. When measuring perimeters from maps, the minimum length of measurement will be determined by the scale of data resolution. Boundary complexity not resolved by the available data will not be revealed and hence cannot contribute to boundary calculations.

Effects on Population Dynamics

In Chapter 3, we examined the factors that affect the changes in numbers for populations that are isolated or semi-isolated. As conservationists and resource managers, we clearly are concerned about keeping local populations from declining to zero, and if they do, we want those empty patches to be quickly recolonized. On the other hand, we may want to encourage the numbers of pests or exotic species to decline to zero. This chapter addresses the role of matrix, if any, in those dynamics. As we will document, the matrix can have multiple influences on events and processes within a fragment or patch. As previously emphasized (Chapter 3), the matrix plays a primary role in determining the quantity of successful movements among habitat patches. This alone means that the matrix is a major player influencing the fate of metapopulations. Two examples will remind us of this basic truism. The percent of British woodlots inhabited by the dormouse *Muscardinus avellanarius*, regardless of size, drops quickly as the distance to the next nearest woodlot increases beyond 800 meters, or 875 yards (Bright and Morris 1996). In south-central Swe-

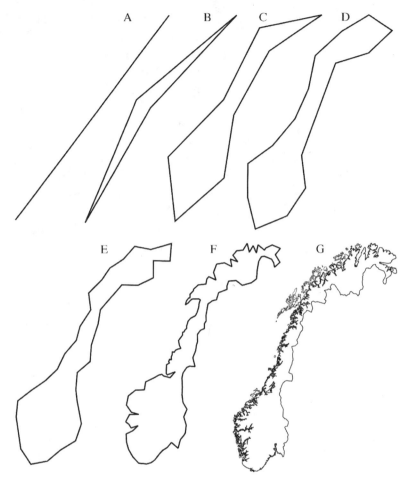

Figure 5.5. Representations of Norway's perimeter (from A to G) become increasingly complex, longer, and more accurate as the measuring segment used becomes shorter. Starting with the longest axis of the country (1,784 kilometers, or 1,115 miles), which is the longest possible measuring unit, A gives such a poor representation of the country that 62 percent of the line actually runs through Sweden and Finland. The line touches the Norwegian border in only four places. In the five representations of B–F, the measurement unit length is progressively halved. By the fourth halving (E), Norway is clearly recognizable and the measured perimeter is 2.3 times greater than in A. With further halving (to 1/32 of the starting length), we begin to see a suggestion of the numerous fjords and islands that characterize the western and northern coasts of the country (F). In G, the actual perimeter is shown in as much detail as possible given the scale of data resolution on the original map.

den, hazel grouse (*Bonasus bonasia*) occur in forest fragments surrounded by a matrix of agricultural lands only if the fragment is less than 100 meters from continuous forest. On the other hand, fragments surrounded by managed coniferous forest (nonhabitat) could be inhabited if they are as much as 2 kilometers (1.24 miles) distant from a source population (Åberg et al. 1995). There are, however, several other important ways the matrix is relevant to population dynamics. One of those is the extent to which the matrix includes dispersal corridors, and that will be discussed in more detail in Chapter 6.

Matrix as a Resource

Whereas a species' habitat may be largely confined to a particular community type, there may be situations in which matrix communities are used for something other than for traveling to another habitat patch. One possibility is that there may be occasional opportunities for abundant food resources in the matrix that can be exploited accordingly. An example is provided by the American marten (*Martes americana*), which in the Sierra Nevada of California requires old-growth forests for its existence. Nevertheless, it prefers forests within 60 meters (197 feet) of meadows, where it regularly exploits dense populations of *Microtus*. It rarely will penetrate more than 10 meters into open meadows but readily uses lodgepole pine (*Pinus contorta*) riparian areas with abundant herbaceous ground cover (Spencer et al. 1983). Meadow voles (*Microtus pennsylvanicus*) do something similar. When artificial patches of grassland are created by mowing, these voles prefer to live on the edges of patches, presumably because that allows them to forage into the mowed areas to take advantage of tender new grass sprouts stimulated by the mowing (Bowers and Dooley 1999). The case of bears making heavy use of salmon on a limited seasonal basis has already been mentioned.

The matrix may also offer access to some resource that is rarely needed by a species, perhaps seasonally, such as a forage plant rich in a scarce mineral, a favorable hibernation site, or access to a pollinator. Ricketts (2004) has shown that tropical forest fragments in proximity to coffee plantations (in Costa Rica) enhance pollinator activity in the coffee, especially within 100 meters of the edge. Edges between different communities may also be especially productive and consequently be attractive places for organisms to live in or near (see preceding discussion).

Finally, we know that some species change habitats according to the season, making ready access to both habitats essential for survival. Rodents may move into the floodplain of a river or lake as it dries out, and then retract to higher ground when high-water conditions prevail (Sheppe 1972, Tast 1966). The endangered salt marsh harvest mouse (*Reithrodontomys raviventris*) lives in the salt marshes of San Francisco Bay. But when the highest spring tides occur, it needs some adjacent upland to retreat to until the water levels decline. Norwegian lemmings (*Lemmus lemmus*) breed in alpine tundra but move into subalpine brush for winter (Kalela et al. 1961, 1971). There the snow is supported by the brush, providing more subnivean spaces for foraging. In temperate climates, commensal house mice (*Mus musculus*) live in agricultural fields in the summer and move to barns and houses for the winter. Rice rats (*Oryzomys palustris*) living in the Texas coastal salt marshes use upland areas as refuge from high tides, for limited foraging in the winter, and as a hideout for juveniles (Kruchek 2004). Many species perform altitudinal migrations between summer and winter ranges, as does the mule deer (*Odocoileus hemionus*) in California. Of course, there are long-distance migrants, especially among birds, bats, and ungulates, and including the monarch butterfly (*Danaus plexippus*). In those cases, however, the concept of patch-matrix interactions is not applicable, as the two or more required habitats are not adjacent.

Finally, there are cases in which organisms completely change habitats in the course of their ontogeny. Many amphibians, for example, need an aquatic habitat for reproduction and growth of larvae and then switch to a terrestrial habitat as metamorphosed juveniles. Bluegill sunfish (*Lepomis macrochirus*) live as juveniles in the vegetated littoral zone of ponds and then move into open water as they become adults.

Juxtapositions of particular combinations of habitats are therefore often important for species, and failure to accommodate this in conservation planning can lead to extinctions or reduced population viabilities.

Matrix as Secondary Habitat

So far, we have considered situations in which the matrix was more or less favorable to dispersers but was not suitable for residency. It is possible, however, that the matrix may serve as secondary habitat. That is, it may be able to support resident individuals of a focal species, although at lower numbers or only intermittently. The reduced densities may be caused by higher mortality rates, reduced reproductive output, or both. Such a population may more easily become extinct and need to be replaced by immigrants from the good habitat patches. This situation has been called a *source-sink axis* (Pulliam 1988, Pulliam and Danielson 1991). A source population is one in which reproduction is adequate to balance mortality and usually to export surplus individuals, as well. Such a population supplies the residents for a secondary habitat. Sink populations are those living in secondary habitat. Their reproductive output is generally insufficient to maintain the population in most years, and hence its continued existence depends on input of immigrants from source patches.

Occasionally, sink populations may produce enough offspring that a few will succeed in immigrating into a source population. This back dispersal may have significant genetic consequences for the recipient populations but is rarely demographically meaningful. There is one circumstance, however, when dispersal from sink to source is critically important. If the source population should unexpectedly become extinct, such as after a severe density crash following a peak in numbers, if a predator or disease should discover the patch and decimate the population, or if it is wiped out by a weather-based catastrophe, the patch can be recolonized from the survivors in the sink habitat. A documented case of this happening in the bank vole (*Clethrionomys glareolus*) has been reported by Evans (1942). Another example is the Alabama beach mouse (*Peromyscus polionotus ammobates*), whose optimal habitat is on the sand dunes immediately inland from the coastal beaches of the Fort Morgan Peninsula. A few individuals live in more inland scrub (marginal) habitat. Following a hurricane, the optimal habitat can be completely destroyed by storm waves, leaving the residents of the interior scrub the only survivors available to recolonize the optimal dunes when that habitat is restored (B. J. Danielson, pers. comm., 2005).

The existence of source-sink dynamics raises a practical caution for land managers. Because population densities in the sink habitat may be as high, or almost as high, as in the source patches (Van Horne 1983), tragic mistakes could be made if density alone is used as a measure of habitat suitability. A manager might conclude that the source patch is not needed because of the large amount of secondary habitat containing good numbers of the target species. Destruction of the source patch would then lead to loss of all of the sink populations, because their long-term persistence was dependent on dispersers from the source population. It is essential, therefore, to ascertain whether populations in secondary habitat are in fact self-sustaining—that is, not a sink—before management decisions of that sort are made. M. T. Murphy (2001) describes an interesting case involving the eastern kingbird (*Tyrannus tyrannus*) in central New York. There the best habitat (riparian) serves as a source only intermittently. The bulk of the population nests in sink habitat (uplands and floodplains), which is only marginally poorer than the source. Reproduction in the sinks contributes importantly both by providing immigrants to the source and by helping to sustain the population in the face of slow long-term decline.

Still another way that matrix as secondary habitat can influence population dynamics has to do with its spatial extent or amount relative to the extent of high-quality habitat patches (Andrén 1994). This ratio of optimal to marginal (secondary) habitat patch area (ROMPA) has been postulated to be a major factor in determining the pattern of population dynamics exhibited by vole (*Microtus*) populations (Delattre et al. 1999, Lidicker 1988a, 2000, Ostfeld 1992; Figure 5.6). The idea is that at low ratios (little optimal habitat) numbers of voles remain low, as reproduction in the favorable sites is just sufficient to maintain modest numbers within the patches and to provide enough dispersers to keep the widely spaced good patches connected. At high ratios, where optimal habitat predominates in the landscape, overall densities will be high, since each patch is very productive and connectivity among patches is high. Such populations will likely show annual changes in numbers corresponding to in-

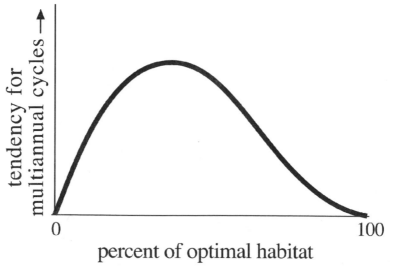

Figure 5.6. Hypothesized relationship between the ratio of optimal to marginal patch areas (ROMPA), here expressed as a percent of optimal habitat in the landscape (*x* axis), and the tendency of voles to show multiannual cycles in abundance (*y* axis) (Lidicker 1992).

creases during the breeding season and decreases during nonbreeding periods. Overall, they will average high densities. At intermediate values of ROMPA, multiannual cycles in abundance will likely be produced. At low-density periods, voles will be restricted to the good habitat patches. As reproduction commences, dispersers will be produced, as voles generally are presaturation dispersers (Chapter 3). These dispersers will find any empty good patches and begin to colonize the matrix of secondary habitat, as well. By the end of the breeding season, and perhaps even the next one, densities will remain low in all of the landscape, as successful reproduction in the optimal patches will have been funneled largely into dispersal. As the matrix and good patches begin to fill up in subsequent breeding seasons, densities will start to increase, producing moderate numbers, a prepeak high. The landscape would then be poised so that in the following year densities would reach peak or outbreak proportions. Such peaks are typically followed by a crash to low numbers, with survival restricted to the optimal areas. In this way multiannual cycles in abundance are generated. Of course, it is more complicated than this (Lidicker 1988a, 2000), but this scenario illustrates how the matrix can play a key role in the demographic behavior of these rodents.

Evidence for the ROMPA effect comes mainly from studies on the California vole (*Microtus californicus*) and European common vole (*Microtus arvalis*) (Delattre et al. 1992, 1996, 1999, Lidicker 1988a, 1992, Ostfeld 1992). It is possible that ROMPA-like behavior is limited to voles, but that needs to be investigated further. In any case, the phenomenon is of some general interest because voles tend to be keystone species in many temperate, boreal, and arctic landscapes, where they form the basis of the predator food chain.

Matrix as a Sink or Stopper

If a matrix is readily permeable to dispersers, much of the reproductive output of a habitat patch may be funneled into emigration. When this output is balanced by immigration, there is no net loss in numbers and many benefits associated with this patchy type of metapopulation will be manifest (Chapter 3). On the other hand, if the matrix is large relative to the size of patches, if it does not support a resident population of the target species, and especially if few other habitat patches are available to dispersers, such a matrix will function as a dispersal sink. Dispersers will be continuously enticed to leave, and there will be little or no immigration back into the patch. Especially in species with presaturation dispersal, the

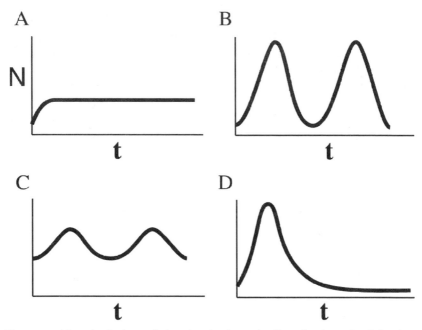

Figure 5.7. Hypothetical population density dynamics for a focal species living in an optimal habitat patch surrounded by four different kinds of matrix habitat. For all four graphs, N is the size of the focal population (*y* axis), and t is time (*x* axis).Matrix is a large sink habitat (A); matrix is a complete barrier to dispersal, effectively isolating the habitat patch (B); matrix is inhabited by a generalist predator that feeds opportunistically in the focal patch (C); focal patch includes a specialist predator on the focal species, but the matrix allows some dispersal of the predator in and out of the patch. Immigration would occur when prey densities are increasing rapidly, and emigration would happen when prey are extremely low in numbers (D).

density of the patch population may be chronically depressed below what it could be if dispersal was more limited (Figure 5.7A).

The matrix as sink is a dangerous trap for dispersers. In an important contribution, Danielson (1991) argues that changes in the proportion of source to sink patches in a landscape (another ROMPA example) can significantly affect the nature of the interaction between two species. For example, if two species are potential competitors, but each does better in the sink habitat of the other, their interactions may change qualitatively as the proportion of sink to source varies. This occurs when species A excludes species B from significant portions of B's sink habitat, thereby facilitating B's search for its own source habitat and overall reducing its losses to

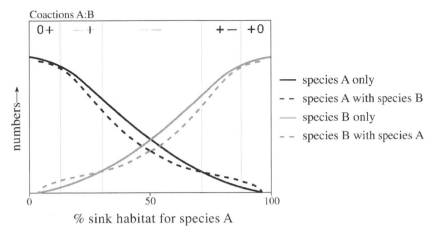

Figure 5.8. Schematic representation based on a model by Danielson (1991, and pers. comm.) of the interactions between two competitor species that share the same habitats. Each species is able to exclude the other from its source patches. However, the sink habitat for each species is the source habitat for the other. The x axis represents the proportion of a landscape occupied by the sink habitat for species A and is the reciprocal of the proportion of its source habitat; source habitat for species A is sink habitat for species B, and vice versa. The y axis is population numbers. For simplicity, the responses of both species are assumed to be symmetrically reciprocal. The resulting coactions (species A:B) are shown along the top bar with o for no interaction effect, + for a positive effect, and − for a negative effect.

the sink. Figure 5.8 illustrates how the interaction coefficient between two such species changes as the landscape shifts from 0 to 100 percent sink for one species, and concomitantly from 100 to 0 percent for the other. Their coaction goes from commensalism to exploitation in favor of the rare species, to competition, and finally to exploitation and commensalism favoring the other species, which has now become the rare one in the landscape.

If, on the other hand, the matrix surrounding a habitat patch is a complete barrier to dispersal (a "stopper"), the patch is effectively isolated from other patches and will suffer accordingly. Dispersal will become frustrated (Lidicker 1975), and numbers will build up within the patch, perhaps reaching high densities (Figure 5.7B). This could produce a chronically abundant population. However, if population growth leads to densities that exceed the patch's carrying capacity for that species, resource damage as well as physiological and behavioral pathologies will generate a likely crash to very low numbers. Such a population will show strongly fluctuating population numbers. The risk is that at the low points of these cycles in numbers, the population will be subject to the negative effects of small populations (Chapter 3). Unless the patch is quite large, extinction of the deme is a likely outcome.

Spillover Exploitation

Not only can patch residents venture out of the patch to use resources in the matrix, but residents of the matrix can penetrate into a patch in search of resources they need, as well. If these invaders happen to be predators or parasites of the target species, they can perpetrate major demographic impacts (Figure 5.7C, D). This phenomenon is called *spillover predation* or *spillover parasitism*, as the case may be.

Much interest and extensive research has focused on the brown-headed cowbird (*Molothrus ater*), a brood parasite. With deforestation, this species has expanded its range and abundance greatly over North America (Rothstein 1994). This cowbird typically feeds in open habitats such as grasslands and prairies. Forest fragmentation has allowed it to come into contact with many potential host species that have had little or no experience with it. It is now known to use over one hundred species of hosts (Kilner et al. 2004). Some of these appear to be seriously impacted, as they have no defense against the cowbird, such as egg or nestling rejection (Lorenzana and Sealy 1999, Peer and Sealy 2004, Robinson, Rothstein, et al. 1995). Since cowbirds are not forest adapted, they effectively parasitize hosts only along the margins of forest fragments (Robinson, Thompson, et al. 1995). Their potential impact is therefore inversely related to patch size, or more specifically to the edge-to-area ratio of the patch. The greater the ratio, the more impact these parasites have on host populations. Desperate measures, such as control of cowbird numbers, have been recommended and instigated in numerous situations where demographic impacts on host species of conservation concern are occurring. A longer-term solution would be to decrease the edge-to-area ratio of forest fragments.

Spillover predation can occur when a predator forages outside of its usual habitat to take advantage of a prey resource, perhaps only occasionally available in sufficient quantity to make such adventures profitable. Alternatively, such predation can happen opportunistically when predators are traveling from one prey patch to another (Figure 5.9). Under such circumstances, a predator can depress prey living in habitats in which it does not habitually forage or in habitats insufficiently productive to support

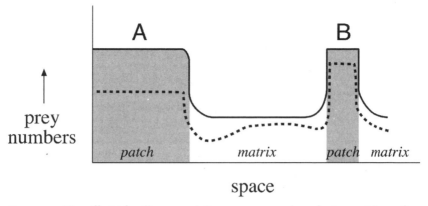

Figure 5.9. The effect of spillover predation on prey numbers for two patches of good habitat (shaded zones) and a surrounding matrix of poor-quality habitat (not shaded). The solid line is the prey densities expected in the absence of predation; it is much lower in the matrix. The dashed line indicates that prey densities are depressed with a resident predator living in patch A. It also reflects the influence of spillover predation both on the matrix and on patch B, which is of the same quality as patch A but by itself is too small to support resident predators. In this scenario, the predators regularly forage out of their home patch, depressing prey numbers immediately adjacent to the good patches, and find small sources of prey that are not too far away (patch B). In the course of these excursions, the predators opportunistically depress prey in the matrix and in small patches of good habitat (dashed line).

resident predators (Figure 3.8). Collectively, spillover predation implies that the impact of predation on a subject population will depend not only on predator-prey dynamics within a patch, but also on the kind of communities in the adjacent matrix, along with the predators associated with it. This is not a rare phenomenon, as examples abound.

In Swedish forest fragments, the impact of nest predation on forest birds by six species of corvids found primarily in open country is dependent on the availability of forest edge (Andrén 1992). Domestic cats depress densities of common voles (*Microtus arvalis*) within 300 to 400 meters (984 to 1,312 feet) of human habitations (Delattre et al. 1999). In Norwegian alpine tundra, weasels (*Mustela nivalis*) and stoats (*M. erminea*) feed on small mammal prey living in low-productivity habitats that by themselves would not support weasel populations (Oksanen and Schneider 1995). Predators living in woodlots in France forage opportunistically into meadows and croplands and thereby reduce the frequency and intensity of vole outbreaks (Delattre et al. 1992, 1996, 1999). Crops close to

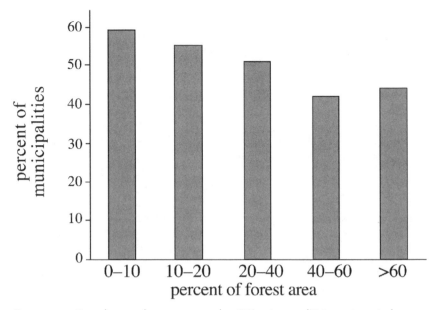

Figure 5.10. Crop damage by common voles (*Microtus arvalis*) in east-central France. The percent of municipalities reporting severe damage is plotted on the *y* axis and is related to the area covered by forest in the landscape of each municipality (*x* axis). (Redrawn from Lidicker 1995b; data from Delattre et al. 1992.)

such woodlots therefore suffer much less damage from voles than crops that are not near woodlots (Figure 5.10). Oksanen et al. (1992) point out that if the proportion of the landscape that is productive enough to support resident predators declines substantially, the total number of predators in the system will be too small to have more than a negligible effect.

Although less investigated, it is plausible that the impact of spillover predation will be influenced by whether the predators involved are specialists or generalists (Figure 5.7). Generalist predators are thought to respond to high densities of prey by focusing their attention on such resource bonanzas (Hanski et al. 1991). When prey numbers drop, however, they lose interest and look for easier pickings elsewhere. Predators of this sort are also likely to be the ones that move widely across landscapes in search of feeding opportunities. Generalists therefore tend to have a stabilizing influence on prey dynamics (e.g., Delattre et al. 1999). They attack the prey highs and ignore the lows. If an isolated prey habitat patch is surrounded by a matrix that prevents dispersal but contains a generalist predator, prey numbers may be prevented from undergoing the boom-

and-bust pattern that one would expect without predation (Figure 5.7). In such a case, predation may actually help an isolated population to persist by effectively preventing population crashes.

Specialist predators are less likely to engage in spillover predation, as they will usually have the same habitat patch as their preferred prey. Within the patch they may generate classic predator-prey cycles in abundance. Kareiva (1987) provides an excellent example involving a beetle predator (*Coccinella septempunctata*) feeding on an aphid (*Uroleucon nigrotuberculatum*) in patches of goldenrod (*Solidago canadensis*). Unless the patch is quite large, both predator and prey risk extinction in the course of strong population fluctuations. Specialist predators might forage in the matrix surrounding their home patch during low phases of prey abundance within the patch. If they can succeed in utilizing the matrix successfully, it will allow the predators to maintain their numbers and put even more severe pressure on the low densities of preferred prey within the patch.

An interesting variant of spillover predation occurs when it is the prey that spill over into the matrix and in that way become vulnerable to predators that are not present or not effective in their home patches. Sakai and Noon (1997) provide an example of this in which dusky-footed woodrats (*Neotoma fuscipes*) living at high densities in shrubfields (northwest California) readily made excursions into adjacent old-growth forest to exploit canopy openings and other nearby shrubfields. In doing this, they sustain heavy predation within the forest, especially from spotted owls (*Strix occidentalis*).

Habitat for Exotics

Introduced, or nonnative, organisms are typically associated with human-generated or human-modified ecosystems. This association is thought to result from one or more of the following factors: (1) similarity between the anthropogenically altered ecosystems and those of the introduced species' native environment; (2) lack of resistance by native species to exotic invasions; and (3) the absence of the invader's usual suite of predators and parasites. Moreover, many successful exotics have "weedy" life history characteristics (Baker 1965) that adapt them for rapid dispersal and invasion of early successional stages in their homeland communities. Since agricultural communities, forest clear-cuts, road edges, suburbia, and other human-modified landscapes often simulate early successional stages, they are suitable habitats for many exotics. It is also increasingly apparent that even lightly disturbed natural communities are less resistant to invasions than intact communities (Chapin et al. 1997). Urbanized areas often suffer numerous introductions of exotics simply because that is where humans purposefully or inadvertently release them. For example, Cohen and Carlson (1998) estimate that between 1961 and 1995, one new species became established in the San Francisco Bay estuary every fourteen days (twenty-six species per year).

Whatever the particular reasons for the success of various introduced species, it is clear that native community fragments surrounded by a matrix of human-generated communities are confronted with an array of exotics. These newly arrived species can have an impact on native communities in a number of ways. First of all, they can affect the nature of the matrix so that its permeability to dispersers from native patches is changed, making it either easier or more difficult to traverse. There are many possible mechanisms for this. Exotics may change the cover characteristics of the matrix, provide a food resource for travelers, diminish a food resource, or change the mix of predators and parasites that will confront dispersers. They can also invade the edges of native community patches. Sometimes this edge penetration can be very deep. For example, W. F. Laurance et al. (1997) document the presence of disturbance-adapted species in central Amazonian rain forest fragments up to 250 meters (820 feet). If the impact of these edge invaders is negative for the native community, that will significantly decrease the effective size of the community patch. If the patch is small, it will change the entire patch into edge habitat, threatening the continued existence of the original community type (Figure 5.3). Particularly ominous, exotics can modify disturbance regimes such as fire intensity and frequency (Mack and D'Antonio 1998), and in that way have major impacts on the landscape.

New encounters between exotics and natives range from benign to lethal. Exotics may be competitors with native species, predators or parasites, and sometimes may serve as novel resources of food or cover that benefit the natives. However, what may be beneficial to one native species may be a disaster to another. For example, improving the lot of one species could transform it into a more effective competitor, and thus harm other natives. Annual grasses, mostly of Mediterranean origins, have massively invaded California grasslands (Hueneke and Mooney 1989). Herbivores such as voles and pocket gophers (*Thomomys* spp.) have profited from this as they often reach higher densities in grasslands dominated by these exotic species (Howard and Childs 1959, Lidicker 1989, M. R. Stromberg and Griffin 1996). However, native perennial grasses and some forbs are correspondingly decreased. Many other consumer species in the community that depend on native plants for their survival will be likewise threatened. The overall score card for exotics is definitely negative. Native species become extinct or threatened, and overall biodiversity declines. The estimated negative economic effects of introductions are huge for crop and range lands alone (Pimentel 2002), but that is a topic beyond the scope of this book. We will return to a consideration of exotics in Chapter 6 where we will consider them further in the context of landscape corridors.

In summary, we have shown that the nature and spatial extent of communities surrounding focal habitat patches, that is the matrix, are critically important components of landscape structure and function. This matrix manifests diverse influences on focal populations and communities. Metapopulation or metacommunity modeling that overly simplifies this dynamic will be correspondingly compromised. Both natural and artificial barriers that are imbedded in the matrix can also be critically influential as they affect movements of organisms and abiotic fluxes. Potentially, there are also long-term evolutionary consequences imposed by matrix features that have not been mentioned explicitly in this chapter. They are discussed, however, in chapters 3 and 6.

CHAPTER 6

Potential Pitfalls of Linking Landscapes

Throughout this book we have emphasized the need, indeed the necessity, of maintaining and restoring connectivity for the increasingly fragmented landscapes on this planet. This imperative supports not only the conservation of biodiversity but also nothing less than the sustainability of the human life-support system. In this chapter, we will explore the possibility that pursuing this essential endeavor can carry the risk of negative effects. Our stewardship must therefore be sufficiently sophisticated that we anticipate and avoid or mitigate unfortunate results that nullify or even worsen what we are trying to achieve in our conservation and planning programs.

As we have stressed, connectivity has to do with the successful transport of organisms or their gametes among patches of the landscape in which the species can live and reproduce. Such movements counter the inevitable march toward extinction that accompanies increasing isolation of habitat patches. Connectivity is determined (1) by the distances between patches relative to an organism's ability to travel, (2) by the quality of the patches themselves, (3) by the hospitality of the matrix that must be traversed, and (4) by the presence of corridors or paths of less resistance for these movements. It is generally the case that in anthropogenically modified landscapes, efforts to maintain or restore connectivity focus on corridors as the most feasible approach. Often it is too late to make meaningful modifications of fragment size and spacing or to change the matrix in any fundamental way. Attention to improving patch quality could be helpful, however, especially in conjunction with corridors, and in cases where the matrix is at least partially permeable to organism transport. Improved quality will serve to increase population size and/or health within the patch, thereby reducing the short-term risk of extinction and generating a larger supply of dispersers.

In this chapter, we consider various factors that may cause corridors to be unsuccessful or to generate unanticipated negative consequences. In Chapter 4 we have already addressed the multifaceted nature of corridors, the importance of spatial scale as it influences corridor properties and function, and the potential benefits of corridors to wildlife and humans. We need only to remind ourselves of a few essential points. The first concerns the essence of what a corridor is, or what it is that makes a specific landscape element qualify as a corridor. We have seen (Chapter 4) that there are many definitions of corridor, but the definitions can be sorted into those that stress structure and those based on function (Hess and Fischer 2001). Definitions that stress structure emphasize linearity in shape, physical connections between patches of the same community type, and physiognomic distinctiveness from the adjacent matrix. Functional definitions focus instead on whether presumptive corridors actually serve as a conduit for the movement of organisms. We think the promotion of movement, rather than whether a corridor shows up well in aerial photos, is the critically important criterion. To qualify as a corridor, a particular landscape element must enhance movement of organisms beyond what is possible through adjacent matrix, or it must permit organisms to cross some barrier.

It is not always necessary for a corridor to be linear. Generally included in the concept of a corridor is the "stepping stone" model of habitat patches that enhance connectivity. In that model, a series of isolated patches of suitable habitat, often only for temporary occupancy, are located so that relatively mobile organisms can move in steps from one survival patch to another (Figure 4.1). The classic example of this is the provision of refuges at appropriate intervals for migrating waterfowl. Date et al. (1991) describe another example involving frugivorous pigeons in northern New South Wales that successfully moved among rain forest fragments and even utilized exotic fruits along the way. Such stepping stones are appropriately viewed as corridors because they function as conduits, although admittedly only for a subset of the species in a metacommunity. Even corridors that are linear may have gaps of various dimensions, making them more like stepping stones.

According to our functional definition, it follows that not all linear landscape features are necessarily corridors, either. Human linear constructions such as windbreaks, roads, canals, riparian buffer zones, and greenbelts are likely intended to serve other functions, such as providing for the movement of people and goods, aesthetics, or recreation or filtering abiotic flows such as wind, pesticides, and fertilizer runoff. While such linear constructs may incidentally serve as wildlife corridors in addition to their primary functions, it is not appropriate to assume that they will inevitably function as corridors, as is sometimes suggested (Hess and Fischer 2001). Because these human-built linear landscape features are often quite different from natural habitats in structure, configuration, and size, it is especially likely that they may promote various negative effects (Box 6.1). In evaluating these features as corridors, it is important to start with a clear understanding of their intended functions.

A newly emerging issue concerns whether the fate of a metapopulation or metacommunity is influenced by the number and distribution of the

Box 6.1.

Potential Disadvantages of Corridors or Causes of Failure

1. Corridors dominated by edge effects
   a. avoidance by potential dispersers
   b. adverse abiotic conditions
   c. increased risk of predation
   d. increased risk of parasitism or disease
   e. competition from exotic species
   f. noise and artificial lighting from adjacent matrix
2. Community drift
   a. species-specific corridor efficacy
   b. inappropriate habitat, length, width
   c. corridor gaps and bottlenecks
   d. sex and age filtering of target species
   e. inadequate assessment of dispersal distances
   f. loss of mutualists or food species
3. Invasions of exotic species
   a. from matrix
   b. from other patches
4. Invasions of deleterious native species
   a. predators
   b. parasites and disease organisms
   c. competitors

5. Demographic impacts
   a. unusual metapopulation processes
   b. synchrony among patches
   c. corridors as dispersal sinks
   d. increased exposure to human depredations
6. Social impacts
   a. corridor inadequate for dispersal of social group
   b. inhibition of movements by territorial residents (social fence effect)
7. Genetic impacts
   a. outbreeding depression
   b. loss of local adaptation
   c. hybridization between taxonomic units
8. Conflicting scientific objectives
   a. habitat versus conduit
   b. needs of target species versus metacommunity maintenance
   c. urgent short-term objectives versus long-term sustainability
   d. alternative conservation strategies
9. Economic impacts
   a. costs of acquisition and construction
   b. costs of maintenance
   c. costs of monitoring
   d. lost opportunity costs
   e. unforeseen negative impacts on adjacent matrix

connections among the various subunits (fragments) that make up the network. Are well-connected units more secure than less-connected ones, and does it matter for survival of the network as a whole? There is increasing evidence (Pletcher 2004) that complex networks, ranging from airline routes and power grids to social systems and genomes (the genetic constitution of an organism) are not connected randomly, as once assumed. In randomly connected networks, the number of connections exhibited by individual units (nodes) is normally distributed about the mean number. Loss of a connection in such an array would be statistically equivalent no matter where it occurred. However, it turns out that in typical networks the number of connections per unit is distributed as a power function; that is, many units have few connections, and a few units have many. The implications of that for us is that loss of a connection to a poorly connected unit (patch) would likely be devastating for that patch but have little immediate impact on the metacommunity as a whole. In contrast, loss of a connection to a well-connected patch might have only a minor immediate effect on that patch, but if such a patch were destabilized by loss of multiple connections, the entire metacommunity would be at risk and would likely disappear. Corridor planners would be well advised to be aware that all connections may not be equal, but rather play various roles in the larger network of which they are part. Henry et al. (1999) make a similar point when they recommend that corridor planning efforts should focus not on single corridors one at a time, but on entire watersheds. In an analogous way, losses of different species cannot be assumed to have equivalent impacts. For example, Zavaleta and Hulvey (2004) emphasize the nonrandom effects of species losses on resistance to invasions in grassland communities.

A final general point about corridors is that in their planning and evaluation we should be clear about what their intended objectives are (Hess and Fischer 2001; Chapter 8). Only then can we evaluate their effectiveness. Objectives generally include (1) the addition of habitat or desired community type and (2) the provision of a conduit for a target species, selected species, or all members of a community type. Sometimes these two objectives are in conflict. For example, if the objective is to move a target species rapidly across a human-created barrier such as a highway, or from one patch to another to minimize predation risk, a corridor must be wide enough with appropriate cover to attract dispersers but not so wide or comfortable that passage is slowed by casual exploration. An ingenious experiment demonstrating this for the tundra vole (*Microtus oeconomus*) was reported by Andreassen, Ims, and Steinset (1996). They found that voles moved readily through a corridor 1 meter (39 inches) wide but slowed significantly if the corridor was 2 meters wide. Similarly, if a corridor contains suitable habitat for a territorial species, residents in the corridor might inhibit dispersers from passing through. On the other hand, sedentary species such as plants or very small animals may take many generations to move from one patch to another. So for them a corridor must contain suitable habitat, or connectivity is not achieved. Of course, plants that rely on animals to facilitate seed dispersal will be able to move along with their animal hosts.

We come now to a consideration of specific things that can go wrong with corridors and cause them to fail in their mission, possibly even making things worse (Boxes 6.1 and 6.2). We hope careful attention to these

Box 6.2.

Grizzlies Ignore Corridors

Canmore, a small town near the entrance to Banff National Park in Alberta, Canada, is home to about thirteen thousand people who generally care about the health of their environment. This gateway community lies in the Bow Valley, which links Banff and Jasper National Parks to the north with wilderness areas in the Rocky Mountains to the south. Under pressure of expanding residential development and golf courses, two corridors were established, one on each side of the valley, in the 1990s. The corridors are 300 meters (984 feet) wide and were designed to channel wildlife, including grizzly bears (*Ursus arctos*), traveling north or south through the Bow Valley. Moreover, this valley is a critical link in the grand Yukon to Yellowstone connectivity scheme (Figure 4.8). A wildlife biologist, Cheryl Chetkiewicz from the University of Alberta, was funded by wildlife organizations and the Alberta government to radio-collar bears to document their movements in the area. The citizens, the provincial government, concerned NGOs, and academics were all cooperating to maintain this vital Rocky Mountain wildlife linkage.

On the weekend of June 4, 2005, a Canmore resident was killed by a bear while jogging along an approved trail (*Canadian Press*, June 12, 2005). This tragedy highlighted a history of bear-and-human conflicts in this area. Of five bears collared by Chetkiewicz, three were removed because of conflicts with people. Moreover, her tracking data clearly showed that bears were spending as much time out of the corridors as in them. She opined that the bears were "not paying attention to our maps."

This fatal incident plus the failure of the bears to restrict their activities to the designated corridors raised many difficult questions for the Canmore citizenry: Are wildlife corridors, especially if used by dangerous predators, compatible with communities like Canmore? Can the corridors be made safer for humans? Are there alternative ways to link the north and south Rocky Mountains? Is it possible to designate and enforce a no-bear zone while somehow maintaining the Bow Valley linkage? Will Canmore residents continue to be willing to live in the presence of large predators? Might bear encounters actually increase if the corridors are eliminated?

For land managers, this case is at the very least a lesson in how the best of intentions to forge linkages through the designation of corridors may not turn out the way they are intended.

potential pitfalls will allow planners to avoid them, or at least to quickly recognize problems when they appear and move to rectify them. As the conservation community gains experience with corridors, there should be fewer and fewer unanticipated surprises. For more detailed discussion of these issues, see A. F. Bennett (1999), Hansson et al. (1995), and Saunders and Hobbs (1991). Early papers calling attention to possible negative influences of corridors include Hobbs (1992), Panetta (1991), Simberloff and Cox (1987), and Simberloff et al. (1992). In evaluating negative impacts, it is important to consider (1) whether these impacts compromise the intended purpose of the corridor; (2) whether they overwhelm the beneficial impacts of the corridor; and (3) whether the benefits and deleterious aspects combined are better or worse than a scenario in which the corridor is absent.

Impacts of Edge Effects

Gosz (1993) reminds us that edge zones (ecotones) can be viewed as ranging over spatial scales from the juxtaposition of biome types (e.g., coniferous forest and grassland) in a regional context to those generated by landscape mosaics of various community types, to edges perceived idiosyncratically by particular species or even individual organisms. Here we adopt the usual definition of ecotone as the product of borders between community types (Chapter 5) and hence focus on that spatial level (Figures 6.1 and 4.3), recognizing that species may differ in their perception of what constitutes a meaningful edge.

Unless corridors are unusually broad, they will likely be dominated by edge effects (Chapter 5, Figure 5.3). Therefore, species that avoid edges will rarely use corridors. Even worse, edge avoiders may be prevented from even finding corridors if, as pointed out in Chapter 5, they refuse to enter the edge zone around their habitat patch. Such so-called "interior species" will be the most difficult to connect to adjacent fragments. Of course, if the distance through the corridor is quite short relative to a species' vagility, even interior species might occasionally jump the gap between fragments when conditions seem the safest or the drive to leave home is the strongest.

Such fervent edge avoiders are not rare. S. G. W. Laurance (2004), for example, has shown that the majority of Amazonian understory bird species, especially insectivores, respond negatively to edges created by an unpaved road less than 40 meters (131 feet) wide. Other research has also

Figure 6.1. Dried vernal pool in the Kesterson National Wildlife Refuge, Merced County, California. As these wetlands desiccate during the summer dry season, successive rings of vegetation are formed. In the center of the pond, the vegetation that is growing where the last remaining water persisted can be seen. Edging the pool is the semiarid brush-grassland typical of this part of the Central Valley. (Photo by William Z. Lidicker Jr., October 1992.)

demonstrated that nonmigratory interior species of birds have great difficulty crossing even narrow gaps in habitat (S. G. Laurance et al. 2004, Diamond 1972, 1973; see Chapter 5). Interior species that are not as mobile as birds might have even more difficulty finding and using corridors. Connectivity for this group of organisms will require broad corridors that include habitats not appreciably modified by edge influences.

Species that do not find edges so inhibiting will be better able to find and use corridors. Even for them, however, successful use of corridors cannot be assumed. Where the matrix through which a corridor passes is heavily modified by humans, travelers through corridors may be confronted with unfamiliar, often exotic, predators, parasites, and competitors (as well as noise and artificial lighting). This is because human-modified communities are often attractive to exotic species (Chapter 5) and may harbor a great many. Often these exotic species, abundant in the matrix, will penetrate some distance into natural communities, which may encompass part or all of a corridor. Artificial corridors may actually be intentionally constructed with exotic species such as *Eucalyptus* trees, often used as windbreaks or for road edges in North America and South Africa. Corridor dispersers will thus face a gauntlet of dangerous species they are not well prepared to confront. It may also be that abiotic conditions in edge-dominated corridors will be unfavorable for dispersers. It is well established that temperature, soil moisture, wind conditions, and even fire susceptibility will be different in edge zones compared to interior conditions.

Corridors as Biotic Filters

One of the most insidious, yet nearly ubiquitous, traits of corridors is that they generally cannot be used equally readily by all members of the communities constituting the joined patches. This differential permeability of corridors produces a filtering effect that may lead to community patches, identical when isolated, drifting apart in community composition. That possibility is not so much an argument that corridors are a mistake as it is a reason corridors may not provide complete community connectivity and therefore may fail to solve all problems associated with fragmentation.

To examine further how this filtering process works, we will begin with two situations in which filtering is not a problem. The first is a situation in which corridors are sufficiently broad that entire communities can move through them. Sedentary species and microbiota may take a number of generations to travel even a hundred meters. For long-lived species such as most trees this means that the process may take many decades. Nevertheless, a wide corridor that is not interrupted by artificial or topographic barriers will provide effective linkages free of biotic drift. Even in that situation, however, one must be cautious about assuming that barriers are absent based on maps alone. Intimate familiarity with the actual corridors (ground truthing) is essential. In some early plans for the Tongass National Forest in Alaska, proposed corridors intended to connect unlogged forest patches were mapped through high mountain passes (W. Lidicker reviewer of forest planning documents). Such corridors clearly would not be usable by most of the forest species, which cannot tolerate the extreme conditions found there.

The second situation where corridors may be completely effective is where large reserves need to be connected primarily for the benefit of a few large species that require very large home ranges or that need to move seasonally between reserves. Species in this category would include top carnivores such as mountain lions (*Puma concolor*) or migratory ungulates. Corridors in these situations could be designed for these target species specifically. This strategy assumes that the rest of the species in the communities involved would not need to be connected, as the individual reserves would be large and varied enough to meet their needs on a sustainable basis. Berger (2004) reviews this strategy for the conservation of migratory ungulates, with particular reference to developing a migratory pathway to accommodate the 225-kilometer (136-mile) migratory trek of pronghorn antelope (*Antilocapra americana*) through the Wind River range in western Wyoming. Van Bohemen (1996) and van Bohemen and Teodorascu (1997) describe examples of how ecoducts have been constructed for the passage of deer over major highways in The Netherlands, and also provide designs for culverts under highways that allow passage of swamp-inhabiting species such as amphibians and otters.

These two scenarios illustrate that successful corridors can range from large swaths of a community type, providing suitable habitat or at least a movement conduit for all the resident species, to a completely artificial and narrow pathway suitable for only a few target species. Most connectivity problems will fall somewhere between these successful extremes. This means that for nearly all corridor planning, some attention should be paid to potential risks inherent in corridors (Box 6.1), whether they are human caused or natural.

The most ubiquitous risk, especially for corridors intended for general community connectivity, is that the nature of the corridor will exclude passage of some species. Patches would thus remain disconnected for some portion of the resident species. Unconnected species will generally suffer a greater risk of local patch extinction combined with a reduced chance for rescue by recolonization. Partially connected communities will therefore gradually change in composition as extinctions without colonization proceed over time. Nearby community fragments may well drift in different directions and eventually fail to resemble each other, as well as becoming different from the original community type that they represented.

Differential use of a corridor by species may result from variations in their ability to find a corridor, their inherent vagility, and their responses to whatever edge effects a corridor might manifest. Since so little is generally known about the travel abilities of most organisms (Koenig et al. 1996), and given the varied type and condition of the proposed corridors, biologists may not be able to estimate correctly the distances over which various species can consistently travel. However, van Langevelde (2000) suggests an empirical technique for estimating the actual threshold distance for connectivity of a given species without having to measure dispersal distances directly. The threshold distance is the distance beyond which the probability of successful crossing by dispersers decreases rapidly. It is determined by finding the distance that best fits an index based on the observed pattern of recolonizations within an array of habitat fragments. This also is not an easy assignment, but it might be useful in some situations.

Not only may a corridor be too long for some species, it may not have an appropriate width, or it could have gaps or narrow bottlenecks that make passage difficult for some species. In experimental landscapes, it has been shown that several species of voles (*Microtus*) will readily cross gaps of mowed grass that are less that 4 meters (13 feet) across, are reluctant to cross those between 4 and 9 meters (30 feet), and only rarely will they travel more than 9 meters across inhospitable matrix (Lidicker 1999). It is also possible that specific corridor attributes could select against certain ages or sexes and hence influence the success of corridor passage. Some corridors may vary seasonally in their attractiveness for travelers, and so favorable periods may or may not coincide with dispersal pulses of different species. Examples of community filtering by corridors are provided by Downes et al. (1997), Forman (1995), and W. F. Laurance (1995).

Finally, species filtering may have more subtle indirect effects. Even species that are well connected may have strong coactions (interactions) with other species that fail to maintain their connectivity. A predator may lose a prey species in this way and vice versa. Various plant species may lose important herbivores or a pollinator species, and competitors may become more or less abundant. Species in mutualistic (cooperative) coactions may lose out if both partners are not similarly connected. A woodpecker that can easily traverse a corridor may become extinct in a patch if a tree species that is important to it fails to maintain connectivity and becomes extinct, or even if the tree changes to a younger age structure and thereby fails to provide what the woodpecker needs. Loss of top carnivores through inadequate connectivity may lead to increasing numbers of medium-sized predators that then put prey species at increased risk (Crooks and Soulé 1999). So it may come to pass that differential species filtering may lead to a cascade of community disruptions that collectively have a much greater impact over the long term than the corridor filtering does in the first place.

In a modeling study, Plotnick and McKinney (1993) conclude that communities will differ in their response to species losses from perturbations, and we can extrapolate this to losses from corridor filtering as well. Specifically, their models suggest that communities in which the component species are strongly connected to each other, both in number of connections and in intensity, will likely experience cascades of extinctions following species loss. On the other hand, the impact on more weakly connected communities that suffer perturbation (or filtering) losses of species may be less likely to include secondary extinction cascades. A corollary of this is Plotnick and McKinney's prediction that species in communities that rarely experience severe perturbations will become strongly connected over time, such as, for example, tropical rain forests or coral reefs. If such a community then endures a severe impact, it may eventually become devastated. In contrast, members of communities subject to frequent perturbations will not develop numerous intimate coactions and as a result will suffer fewer losses when extinctions do occur.

The message for us here is that species losses caused by corridor filtering may sometimes have major consequences for community integrity, and sometimes the impacts will be minor. The risk of drift, of course, needs to be evaluated with respect to the corridor objectives. A general rule for land planners would be this: The larger the patches to be connected and the more generally passable the connecting links, the less likely it is that problems will arise from community drift.

Facilitation of Invasions

Invasions of unwanted species into habitat fragments are perhaps the most widely appreciated of the potential deleterious impacts of fragmentation (Holdaway 1999, Pimentel 2002, Pimentel et al. 2000). It is generally acknowledged that introductions are second only to habitat loss as causal agents in the decline and extinction of species worldwide. This is largely a problem generated by humans, because we are the causal agents for almost all of the introductions of exotic species, either intentionally or accidentally. We have not only transported and released thousands of species outside their natural distributions, but also placed anthropogenically altered communities, in which exotics tend to thrive, immediately adjacent to natural ones. Thus we have set the scene for massive interference in the structure and function of native communities. It is clear that the two most important agents of species endangerment, habitat destruction and introduction of exotics, are themselves connected in a synergistic fashion. Considerable research effort is directed toward helping us better understand this invasive process (D'Antonio et al. 1999, H. A. Mooney and Drake 1986, Rejmánek and Richardson 1996, Vitousek et al. 1996, Zavaleta and Hulvey 2004).

Corridors can contribute to this conundrum of negative impacts in two ways. Most important, they usually provide additional edge habitat for exotics living in the matrix to invade natural habitats (Crooks and Soulé 1999, C. C. Hawkins et al. 1999, Panetta and Hopkins 1991). This happens when the length of the corridor exceeds the combined widths of the corridor where it abuts the patches that are being connected. When invasive species are so situated, they can interfere with the success of native species whether they are living in the corridor or just passing through. They can do this in multiple ways. First of all, exotics can alter the habitat in corridors by changing cover characteristics or food supplies. Other possibilities are increased predator and parasite pressures and enhanced interspecific competition. On a longer time scale, exotics living on edges can be selected for improved capability of invading the native communities and thus gradually extend their colonization of the patches as they adapt to the new conditions.

The second way that corridors can be a problem is by allowing introduced species to spread from one patch to another (Hess 1994, 1996, Watson 1991). If an exotic successfully invades a patch of natural habitat, it may then spread to additional patches aided by corridors. Johnson and Cully (2004) describe a likely example of this. They report that colonies of black-tailed prairie dogs (*Cynomys ludovicianus*) that are connected by dry drainage channels, much used by these rodents for dispersal among colonies, are more likely to suffer heavy mortality from sylvatic plague (*Yersinia pestis*) than colonies not so connected. Their data come from five national grasslands and cover four years.

Less appreciated is the role that corridors may play in expanding the ranges of native species. Sometimes this entails regional or local range extensions, a process that corridors are often designed to achieve. More commonly, it involves colonization of a patch by native species previously absent from that patch. Such events are mostly not a problem, as the invaders are usually normal components of the community type being colonized. They are adapted to patch conditions, and the resident species are adapted to them. Conservation problems do arise occasionally, however, particularly if a target species of concern is negatively affected by the arrival of a new predator, parasite, or competitor. In the normal scheme of things, habitat patches that are isolated, either permanently or periodically, may lose species of predators or parasites that are particularly at risk because of the isolation. They may be lost by chance (demographic stochasticity) or by genetic deterioration (Chapter 3), especially if the patch is small. Or their prey or host species may become extinct or too rare to support their continued existence in the patch. Prey or host populations may then thrive in the absence of these top consumers. If recolonization of the predator or parasite should occur subsequently, deleterious impacts on prey or host will ensue. This dynamic pattern is ordinarily not a problem and in fact is one of the processes that is intentionally abetted by connecting patches. Only if a species of conservation concern is so reduced that it survives in only one or a few patches will colonization by a native predator or parasite be potentially disastrous. A poorly conceived corridor construction project may be a bad idea in such a situation.

In assessing specific landscape configurations with the objective of either encouraging or discouraging the persistence of various carnivores, it is important to keep in mind that different predators perceive a patchy environment at different spatial scales (Gehring and Swihart 2003). Large and especially more vagile species are better equipped to traverse multiple patches and often are more adept at crossing inhospitable matrix. For them, corridors composed of the same or a similar community type as that of the preferred patch type are not so important. Instead, they may need corridors that allow them to cross human-made barriers. On the other hand, small and less mobile carnivores tend to be tied more closely to specific habitat features and may therefore benefit from corridors that possess the necessary attributes. Orrock and Damchen (2005) report that in patches of clear-cuts imbedded in loblolly pine plantations (South Carolina), two species of plants experience heavier seed predation by rodents in connected than in isolated patches of the same size. Planners contemplating adding corridors to a landscape should consider how the various predators potentially present in the area would be affected by the improved connectivity. Effects could come from changes in the connectivity of the predators or indirectly through resulting changes in their prey base.

An active area of research addresses how patch or host connectivity influences disease and parasite incidence and virulence (Hess 1996, Gog et al. 2002, McCallum and Dobson 2002). In Chapter 3, we mentioned how human measles infections can persist chronically only in populations larger than 250,000 people (Grenfell and Harwood 1997, Keeling and Grenfell 1997). In smaller populations, measles becomes extinct, although it can, of course, recolonize. Ecologists have generalized this phenomenon in the form of *refuge theory* (B. A. Hawkins et al. 1993), which posits that the ability of a parasite (or parasitoid) to reduce host population numbers is inversely related to the proportion of hosts that are unavailable to the parasite. Unavailability could be in the form of genetic or acquired immunity, or it could result from living in an isolated habitat patch. Connecting such refugial patches could destroy the patches as refuges and increase their vulnerability to parasites (Hess 1996). These same principles can be applied to predators. Small patches may not be large enough to support a predator population, or they may be too isolated for predators to find them. Prey densities may actually thus be higher in such refuges. There may, however, be a compensatory interaction with parasites and diseases, in that higher prey numbers permitted in the absence of predators could lead to increased parasite or disease transmission within isolated patches if a pathogen should reach them. Moreover, increased stress from high population numbers could increase the susceptibility of hosts to the pathogens.

Modeling studies by Gog et al. (2002) and McCallum and Dobson (2002) argue that isolation is a risky strategy, and that connectivity actually improves the chances of metapopulation survival. Their models incorporate the realistic possibility of resistance to a pathogen developing so that infected patches do not invariably become extinct, as assumed by Hess (1996). They also include alternative or reservoir hosts in their systems. Both papers conclude that while connectivity encourages the spread of pathogens, the benefits of enhanced recolonization of empty patches and alleviation of risks associated with small population sizes (Chapter 3) outweigh the costs of disease spread. There are situations, however, when a target species will be at serious risk of metapopulation extinction, such as where patch connectivity leads to a higher rate of patch extinction for the target species than for the secondary host.

When multiple hosts for the pathogen are involved, interesting indirect effects might occur. For example, it is known from studies in both France and China that conditions conducive to supporting high numbers of rodents (especially of the subfamily Arvicolinae, i.e., voles) are associated with relatively high incidences of alveolar echinococcosis in humans (Giraudoux et al. 1996). This pathology is caused by larvae of the tapeworm *Echinococcus multilocularis* infecting the liver and is presumably transmitted to humans via dogs. In other words, landscape arrays that encourage high rodent densities affect the tapeworm's secondary host, namely humans in this case.

A final point concerns whether changing connectivity through corridors might influence the virulence of parasites and diseases affecting target species. Not much is known about this, and there may well be confounding factors in any given instance. Nevertheless, one provocative report (Herre 1993) suggests that increased opportunities for transmission lead to increasing parasite virulence, especially if transmission is between unrelated individuals. The study involved nematode parasites infecting fig wasps in Panama. It concluded that wasps characterized by the most subdivided population structure had the most benign parasites. It is too early to know how general this result is, but it does raise the possibility that increased virulence may in time result from increasing connectivity of metacommunities. Some additional studies support this possibility (references in Herre 1993, Dobson and Merenlender 1991). If virulence does evolve, host populations may be depressed, and other secondary repercussions in community structure may ensue accordingly. In most cases, however, improving connectivity among fragments will only partly restore the conditions that existed before fragmentation and hence will produce a level of virulence to which a focal species will already be adapted. Problems might arise with new diseases or with species on the edge of extinction.

Demographic Impacts

We have already seen how variations in connectivity among habitat patches may influence population numbers by encouraging or discouraging predators or parasites and diseases. There are, in addition, several other demographic impacts that can be manifested by corridors (Box 6.1).

Unusual Metapopulation Processes

In Chapter 3, we reviewed metapopulation dynamics and emphasized how changes in the level of movements among semi-isolated patches were critically important in determining the long-term fate of metapopulations. In general, improved connectivity leads to better chances of metapopulation persistence. Here we review several situations in which that rule may be violated. Suppose a metapopulation consists of a large number of well-isolated patches, and an undesirable predator, usually an exotic species, enters one patch in the system. This predator could either destabilize a prey species to the point that it would have a high probability of local extinction, or fail to persist in the patch because it was too small or too unproductive. In such a situation, the unwanted predator could not successfully invade the system, because it could not sustain itself in any one patch. Now, if the patches became better connected so that the predator could more readily move among the fragments, it could likely persist in the metapopulation as a whole by regularly abandoning exploited patches and finding those that have recovered. A. D. Taylor (1991) critically reviews thirteen examples where a predator appears to persist in an array of fragments even though it would not be able to do so in individual patches. Only one of these cases, a lizard preying on spiders, involves a vertebrate.

A second example involves the black-tailed prairie dog and plague interaction already mentioned. Stapp et al. (2004) report on observations of colony extinctions caused by plague in a twenty-two-year study in northern Colorado. *Yersinia* is transmitted to the prairie dogs by fleas, so the two important variables in the system are (1) factors that affect flea survival and reproduction, and (2) the movement of infected prairie dogs between discrete colonies. The situation is complex, but two results of the study are both counterintuitive and relevant here. Metapopulation theory (Chapter 3) would predict that small colonies would suffer the highest rate of extinctions. These investigators found that extinctions were most likely in both very small colonies and in the largest colonies; medium-sized colonies had lower than average rates of extinction. The unexpected poor performance of large colonies was apparently caused by the rapid transmission of plague among the many individuals within them. The second surprising finding was that the probability of colony extinction was also strongly related to the size of the nearest neighboring colony one year before. Colonies, regardless of size, that had large neighboring colonies were more likely to go extinct the following year, in this case presumably because of movements of infected prairie dogs into neighboring colonies.

The lesson from these two examples is that we need to be prepared for unintended consequences of reconnecting landscapes, especially when exotic species are potential invaders into a metapopulation structure.

Synchrony among Patches

If all the demes constituting a metapopulation have completely independent demographies, an unlikely situation, the probability of extinction of the entire metapopulation is the product of the extinction probabilities for each deme (e^n) (Chapter 3). Obviously, if the number of demes (n) is large, the chances of metapopulation extinction become vanishingly small. However, as the synchrony among the demes increases, the number of independently fluctuating units declines, effectively, if not actually, reducing n. The probability of metapopulation extinction then approaches that of a single deme (e^n approaches e). Synchrony happens for two reasons: (1) Demes can respond demographically in the same way to extrinsic factors such as seasonal changes or unusual events. (2) Movements of individuals among the demes cause a loss of demographic independence. And, of course, a combination of the two also is possible. Stochastic processes may add an additional element of variability.

This simplistic analysis so far assumes that the probability of extinction of a given deme is a constant, at least over a substantial time period. Actually, however, this probability likely varies, and one reason for this is the changing chance of a patch receiving immigrants. Immigrants can rescue a population that otherwise is headed for extinction. And the timely availability of immigrants to a patch may well depend on neighboring populations being in a different demographic phase. That is, a patch enjoying an increasing density may supply dispersers to an adjacent patch that is at a low density. On the other hand, if all the populations in an area are demographically synchronous, all the patches will have a surplus of individuals or be in need of rescue at the same time, and rescuing will be less likely. Demographic asynchrony therefore increases metapopulation persistence by enhancing the potential for demic rescue. The reality of this process has been supported by modeling (Kallimanis et al. 2005, Stacey et al. 1997) and some empirical data on spotted owls (*Strix occidenatalis*; Lahaye et al. 1994), pikas (*Ochotona princeps*; A. T. Smith and Gilpin 1997), and white-tailed ptarmigan (*Lagopus leucurus*; Martin et al. 2000).

Given that, at least in principle, the probability of metapopulation loss increases with demographic synchrony among the patches, can conservation planners manage so as to reduce synchrony? The most efficacious approach is to achieve a metapopulation structure that is both large with respect to the number of demes and spatially diverse so that many different microhabitats are included. This strategy will make it unlikely that a dangerous level of demographic synchrony will occur. If this strategy is not possible, demes will need to be isolated to the extent that synchrony caused by dispersal can be reduced. This option carries the risk that if isolation is too great, all of the risks associated with disconnected populations will occur (Chapter 3). With large and diverse metapopulations, corridors are not a risk factor, but with small and relatively homogeneous metapopulation arrays, corridors may indeed pose an added risk.

Empirical data for this synchrony effect are scarce, although a few examples have been given. There is, moreover, one experimental study (Burkey 1997) that may provide additional evidence relating to synchronous effects. Burkey studied simple laboratory metacommunities composed of a few species of bacteria and protozoans living in culture media. He established small and large communities ("patches"), and some were connected by corridors, while others were not. The time course to 100 percent extinction of communities was determined. Community size turned out not to make much difference, but the patches connected by corridors were all extinct by sixty-four days, compared to eighty-four days for unconnected patches. There are many reasons why this result cannot be generalized to much more complex natural metacommunities, but it is at least suggestive and adds to the evidence that synchrony among demes may be a factor in metapopulation dynamics. As always we must keep in mind, however, that demic synchrony might have important advantages, as well. Both modeling and empirical evidence demonstrates that synchrony is one way of keeping predator and parasite populations at low levels. So the best course for conservation planning may not always be obvious.

Corridors as Dispersal Sinks

As we have seen (Chapter 5), the matrix community surrounding a habitat fragment may act as a dispersal sink, siphoning off dispersers from the patch and actually keeping the density of a focal species at chronically low levels (Figure 5.6A). This is most likely going to happen if the matrix is extensive, so that it always seems "empty" to a potential disperser, or is relatively attractive secondary habitat. In the latter case, dispersers may become resident in the matrix and attempt to breed. Reproduction, however, will generally be inadequate to replace losses, and the matrix will serve as a "demographic sink," dependent on continued dispersal from the favorable patches for the sustained residency of the focal species (Schlaepfer et al. 2002). Little or no dispersal from the sink back to the good patches will occur because there is no surplus reproductive output. As discussed in Chapter 5, such habitats are not necessarily deleterious for metapopulation persistence, but they can have local impacts that should be evaluated.

Corridors, like matrix, can also function as dispersal sinks, resulting in populations that cannot persist without continued immigration from source populations (Soulé and Gilpin 1991). In fact, they may be even more likely to behave that way because effective corridors are attractive to dispersers from source patches. Especially good corridors may actually provide resident habitat for dispersers. If this habitat is inferior in quality or subject to intensified edge effects in the corridors, it may act as a demographic sink. Dispersers may then be enticed to live in the corridors but end up contributing little or nothing to future generations. For example, high rates of songbird nest predation were observed in remnant floodplain forests surrounded by rice agriculture in the Consumnes River Preserve in California's Central Valley. This predation has been attributed in part to rats (*Rattus*) that are now found in high densities (J. Marty, The Nature Conservancy, pers. comm.). Corridors can also produce demographic sinks if they simply lead dispersers to suboptimal habitat patches. Thus, if a corridor is merely a drain on a patch rather than providing connectivity to other favorable patches, its negative effects may outweigh its benefits.

Although not especially involving corridors, two studies illustrate how attractive habitat along roads can become demographic sinks because of added mortality caused by road traffic. One study (Mumme 1994) concerns Florida scrub jays (*Aphelocoma ultramarina*), and the other (Reijnen and Foppen 1994) is on willow warblers (*Phylloscopus trochilus*). Also, Baranga (1991) has described the sink-like qualities of the Kibale Forest Game Corridor in Uganda.

Increased Exposure to Human Depredations

A potential negative impact in some situations is that corridors might serve to improve access for *Homo sapiens* to wildlife resources. Like exotic predators, humans may find that corridors provide more accessible edge habitat for hunting or allow them to intercept traveling prey more easily. Prey may be further disadvantaged by being confined to a narrow strip of cover, which may offer inadequate protection. Moreover, prey may suffer the disadvantage of traveling outside their normal home ranges, where they are less familiar with safe hiding places and escape routes. Adina Merenlender observed traps for lemurs in the Masoala Peninsula in Madagascar that were specifically set along narrow corridors of forest. This strategy by local hunters significantly increased the chances that lemurs would be caught. The concentration of waterfowl hunting in migratory flyways is also a well-known phenomenon. Again, the report by Baranga (1991) on the Kibale Forest Game Corridor is relevant here, because local people have moved into the corridor and used it to harvest game species.

Social Behavior

Earlier (Chapter 3), we discussed how a species' social system might influence its ability to travel between habitat fragments. Here we ask whether corridors pose any special problems relating to social behavior. As we will see, there are several reasons why a corridor that seems perfectly adequate to accommodate movements of a focal species may nevertheless fail, with social behavior being the culprit.

In species with complex and perhaps obligatory social groupings, successful dispersal may require that groups of individuals travel together. Corridors will need to be able to accommodate group movements. This may involve better cover and food requirements than would be needed by individual travelers, and perhaps special features such as tree cavities that would be suitable for group resting. W. F. Laurance (1990, 1995) describes an example in which a highly social arboreal marsupial (*Hemibelideus lemuroides*) never used corridors. Even though this species was common in nonfragmented rain forest (Queensland, Australia), it never occurred in habitat fragments. Either the available corridors were not perceived as suitable, or cohesive social forces successfully resisted any long-range dispersal behavior.

At the other extreme are species that vigorously defend territories against intruding conspecifics. If a corridor contains usable habitat, even if only of marginal quality, individuals will establish territories in the corridor, and the resulting "social fence" may inhibit the passage of vagrants who just want to pass through. Where this is the case, corridors that function simply as conduits rather than as additional living space could be more effective. More data on this possible role of territoriality are much needed.

Genetic Impacts

The role of fragmentation in influencing the genetic structure of metapopulations has been detailed in Chapter 3. In general, connectivity of demes has favorable genetic consequences. Now we need only consider whether there are possible negative impacts of connecting previously disconnected fragments. Are there genetic risks implicit in reconnecting landscapes?

Potential negative genetic repercussions of corridor construction all relate to interbreeding among individuals from populations that have differentiated from each other. Such differentiation might result from genetic drift or differential selective pressures operating in different places (Chapter 3). Because negative consequences arise from connecting populations with a history of separate evolutionary trajectories, we can anticipate that such consequences will be most likely to occur if populations have been separated for a long time or if large spatial scale connections are developed (Rhymer and Simberloff 1996). For example, corridors that allow organisms to cross major dispersal barriers that have previously served to allow subspecific differentiation within a species could lead to hybridization (introgression) and subsequent loss of those subspecific entities. Even taxa at the species level of differentiation can sometimes be at risk. Rare species are particularly vulnerable to this mode of genetic extinction. Ellstrand and Schierenbeck (2000) make the ominous suggestion that hybridization among plant species can stimulate the evolution of invasiveness.

There are many examples of North American birds in which eastern and western forms have been brought into hybridization as their distributions have met through urbanization of the Great Plains. Towns with planted trees have provided stepping stone corridors for expansion of the ranges of forest-adapted species. Three well-known examples are the yellow-shafted and red-shafted flickers (*Colaptes auratus auratus, C. a. cafer*), Baltimore and Bullock's orioles (*Icterus galbula, I. bullockii*), and rose-breasted and black-headed grosbeaks (*Pheuticus ludovicianus, P. melanocephalus*). In all three of these cases, it appears that the hybrid zone has stabilized, so taxon identity may not be lost. Another example is the possible future demise of the spotted owl, an inhabitant of western North American old-growth coniferous forests, through hybridization with the barred owl (*Strix varia*), an eastern species that invaded the West in the 1960s (Hamer et al. 1994). Introgression was the cause, or at least a contributing factor, in the recent extinction of three species of North American fish (Rhymer and Simberloff 1996). A great many plant taxa have disappeared or are threatened by interspecific introgression, as well. Often it is an exotic species that genetically swamps a native taxon. Human-modified habitat changes may bring formerly separate taxa together or create a hybrid habitat that encourages interbreeding. Still other examples stem from the deliberate introductions by wildlife managers or hunting groups trying to improve game availability. Many more examples of taxonomic disruptions can be found in Rhymer and Simberloff (1996).

Disruption of taxonomic entities as depicted in these examples is normally not planned but is an inadvertent result of human activities that connect habitats on a regional or continental scale. Negative genetic impacts are possible, however, on a much more local scale, as well. Where isolation has allowed selection to move populations toward improved local adaptation, enhancement of gene flow may cause the loss of these evolutionary improvements. This would be an example of outbreeding depression, because mean fitness would decline in the affected demes (Templeton 1986). Such depression could also occur even if local differentiation was caused by drift instead of differential selection. Such differentiation might produce coadapted gene complexes or successful polymorphisms (Chapter 3) that would probably be disrupted with increased genetic exchange between demes.

Although there is clearly a potential for negative genetic impacts with implementation of corridors, the circumstances in which such effects are likely are limited. Particular vulnerabilities are associated with new connections over large spatial scales and where local populations have the potential for local differentiation. Nevertheless, it is important to keep in mind that different species view the world at different spatial and temporal scales, and therefore a corridor that is beneficial for one species may introduce unfortunate genetic impacts to another.

Conflicting Scientific Objectives

Virtually all conservation planning involves compromises. The various stakeholders will generally have different objectives and even different value systems. Successful planning will need to take all of that into account. In subsequent chapters, these conflicts will be addressed in more detail. Here we are concerned not with social, political, and philosophical conflicts but with conflicts over what would be the best scientifically based plan, assuming nothing else matters. As we will see, the pros and cons of corridor preservation, reconstruction, and maintenance often run into conflicts over objectives and priorities even in a strictly scientific context (Box 6.1).

If one is so fortunate as to be confronted with the task of connecting large blocks of relatively pristine habitats, and the possibility exists for preserving or reconstructing broad corridors in which edge effects are minimal, conflicts in objectives will be absent or minimal. The preserved and well-connected landscape will need minimal maintenance. Given such a suitable structure, the communities contained therein could take care of themselves. Unfortunately, that is rarely the way it is. In most cases, planners are faced with making the most of a bad situation. Pristine conditions have vanished, the possibilities for corridors are limited, and ongoing management is required to maintain even the limited objectives that are possible. Almost always there is the added threat of intrusion posed by numerous exotic species, which native communities are not generally able to resist.

Given the realities of most planning situations, there will inevitably be room for argument about priorities, what is possible to achieve, and the best scientifically based tactics to move forward. The bases for most of these conflicts have already been addressed. Should a corridor, for example, be designed as an efficient conduit for guiding movements of organisms, or is it more important to provide a corridor with additional habitat, even if marginal in quality? Should the focus be on connecting target species that are rare, endangered, or threatened or that are keystone species whose loss would significantly alter the communities to be connected? Or is it critically important that corridors offer movement potential for as many species as possible so that entire communities are connected into a meta-community? Then, of course, there are priorities based on urgency. Are there species of conservation concern that require immediate attention, or are there land acquisition priorities that cannot wait? How much attention should be given to the more difficult-to-defend long-term objectives such as predicted genetic consequences of a given plan, evolutionary sustainability, and future adaptability? Should planning take into account global climate change projections and what they herald for the future suitability of current plans?

Even after all of the priorities and differing objectives have been thoroughly analyzed and agreements reached, conflicts may persist. This is because there may be alternative strategies for achieving the same goals. If improved connectivity is a goal, should corridors be constructed, and, if so, how many and where? Would an alternative be to improve the permeability of the matrix communities? Does it make sense to use translocations to move species among community patches (Griffith et al. 1989)? Do alternative plans imply different degrees of participation by the local human communities? Many more issues of alternative conservation strategies are possible. One thing seems certain, and that is that the more degraded a landscape is to begin with, the more difficult it will be to resolve conflicts, the more complex any restoration efforts will be, and the more likely it is that extensive maintenance efforts will be required.

Economic Impacts

The critical importance of economic issues in land-use planning will be discussed in more detail in later chapters. In this chapter we mention them only to complete our litany of potential negative influences emerging from efforts to reconstruct landscapes. In particular, we need to ask whether preservation and development of corridors make sense both biologically and economically. The best plans, if not implemented, are an exercise in futility—although, granted, they may also be a learning experience with potential future benefits.

Here we emphasize potential negative economic impacts (Box 6.1). These, at least in principle, can be relatively easily estimated in monetary terms, although even there unpredictable outcomes are always possible. Pimentel et al. (2000) discuss, for example, economic costs associated with invasions of exotic species in the United States. In any real situation, of course, there are economic benefits to be realized from a conservation plan. These are traditionally ignored in land-use planning. This is partly because benefits are often difficult to convert into the monetary language of economics, and partly because they are typically difficult to predict in quantitative terms. Benefits may be aesthetic, educational, or research oriented; they may involve unspecified improvements in the quality of life, the conservation of biodiversity, recreational opportunities, or improved ecosystem services (Daily 1997, Daily et al. 1997). The latter may include cleaner air; more reliable and cleaner water supplies; protection against soil erosion; degradation of pollutants; mitigation of floods and droughts; pollination of crops; weed and other pest control; and moderation of temperatures, winds, and waves. These are all things that traditional economics is ill equipped to evaluate, although the new wave of ecological economists is making good progress. One attempt to do this concluded that the value of these services per year was almost double the combined gross national products of all the nations on earth (Costanza et al. 1997).

Four Case Studies Analyzed

A number of studies in the published literature conclude that corridors are not helpful in particular situations. Such studies could be used by some to deny the value of corridors in general. Therefore, we provide here a brief analysis of four such studies to illustrate the need to evaluate the generality of their findings. Collectively, they illustrate the important conclusion, which we have also stressed, that corridors are not the answer for rectifying all situations in which communities are fragmented. The studies illustrate diverse situations, including boreal forest, urban parks, pine plantations, and grasslands. For a careful analysis of thirty-two other corridor studies, see Beier and Noss (1998). Their conclusion was that the better-designed investigations supported corridors as indeed being valuable conservation tools.

Hannon and Schmiegelow (2002) studied boreal species of forest birds for five years following clear-cutting in north-central Alberta. Although they found that abundances were higher in large uncut patches, fragments with and without corridors were not consistently different. The conclusion was that corridors did not improve the conservation value of fragments, at least for most boreal bird species. The study had two major limitations. First, it involved only a suite of migratory species, and, second, observations were made immediately following clear-cutting. In retrospect, it seems unsurprising that highly mobile migratory birds would not find corridors useful in finding forest fragments. The short time interval following fragmentation implies that the results are very short term and could not be used to assess an equilibrium or long-term effect.

Fernández-Juricic (2001) studied birds resident in urban parks in Madrid and their use of corridors, represented by the tree-lined streets connecting the parks. There were fourteen species of birds resident in both the parks and corridors during the two-year study. Only six species, however, were usable in their analysis. Of these, five species increased their use of the corridors when they were abundant in the parks, and one did not. The author concluded reasonably that the five species were using the corridors as secondary habitat, and not as conduits for movement between the parks. It seems that this is a good example of the risk of assuming that what may appear structurally to be a corridor actually functions as one. In this case, the evidence from these mobile urban birds is that tree-lined streets are not functioning corridors, but may serve as important secondary habitat.

Danielson and Hubbard (2000) took advantage of clear-cutting in loblolly pine plantations in South Carolina to follow movements of several species of rodents. Some of the clear-cut patches were isolated by a matrix of forest, and some were connected by corridors of cut timber. Marked individuals of two species (*Peromyscus polionotus* and *Sigmodon hispidus*) that are adapted for living in the open habitat created by clearcuts were followed. The average number of individuals of either species did not differ between isolated and connected patches. It is interesting, however, that isolated patches had male-biased sex ratios, implying that males more readily crossed matrix habitat to locate those open areas. The conclusion was that corridors are not especially useful for those two species, a finding not surprising given that the two focal species are excellent dispersers adapted to locating isolated patches of open habitat. In this area, open habitat is often ephemeral, representing early successional stages. The life history pattern of these two rodent species, in fact, indicates that they are unlikely to be dependent on corridors.

The final example (Collinge 2000) involved insect movements among fragments of grassland. The study was done in native mixed prairie in north-central Colorado over a three-year period. Replicated habitat patches were established by mowing and were in three sizes: 1, 10, and 100 square meters (1 square meter = 10.76 square feet). Some plots were connected to uncut prairie by 1-meter-wide (3-foot-wide) corridors, and some were not; 10 meters (33 feet) separated the patches. Insects were sampled for two years with sweep nets, so only species accessible by this procedure were monitored. Corridors seemed to have little influence, although species richness was slightly higher in medium-sized patches with corridors than in those without. In a third year, recolonization was studied, using only the medium-sized fragments, following removal of insects with 360 sweeps of the net in each plot. The time scale of sampling was from one hour after insect removal to fourteen days. By day fourteen, all plots had the same level of recolonization, but isolated plots approached this level more gradually. Corridors slightly increased the recolonization rate of the less vagile species. Clearly, there are limitations to this research that influence the usefulness of the results. One is the extremely limited spatial and temporal scales of the study, and a second is that only relatively vagile insect species would be vulnerable to sweep net sampling.

This diverse group of four investigations emphasize that (1) it is prudent to be cautious about assuming that corridors will inevitably improve connectivity among habitat fragments; (2) what looks like a corridor to us may not be so judged by target species; (3) various life history features may be clues as to the potential usefulness of corridors; (4) results may be dependent on the spatial and temporal scales of the observations; and (5) the general significance of a study may be discerned only from scrutiny of the details.

In summary, this chapter has presented an account of what could conceivably go wrong with efforts to maintain or provide linkages among the various components of our increasingly fragmented world. Potential negative impacts are organized into eight topics, as summarized in Box 6.1, and surely there are things that we have missed (Box 6.2). Singly or collectively, these factors can cause corridor projects to fail or be diminished in their effectiveness. Do they add up to a denial of the importance of our efforts to achieve a better-connected world? They do not. While we have attempted to present these possible difficulties as strongly as possible, we believe that in the vast majority of cases the benefits of a corridor will outweigh the negativities. Nevertheless, it is prudent for all parties involved in such projects to be aware of potential problems so that they can be eliminated, alleviated, or repaired through design changes or ongoing management (Chapters 7, 8, 9). In all cases, project evaluations should have the best possible information available for calculating the costs and benefits of any proposal, and especially to be able to compare their value with those of alternative plans or, most importantly, with no plan at all.

Corridor Design, Planning, and Implementation

Part III aims to bring the scientific issues discussed in Parts I and II into an applied context and offer guidelines on corridor design, planning, and implementation. Here, we provide a thorough examination of factors to consider when working on linking landscapes for various conservation objectives.

Chapter 7 examines corridor design considerations related to focal species and habitat attributes. We also discuss overall corridor quality, including the importance of continuity, composition, and dimension, as well as the landscape context. The discussion in Chapter 8 is devoted to the importance of planning for connectivity, beginning with assembling a team, deciding on the appropriate scale, and identifying corridors for conservation and restoration. We describe various available approaches and tools to help facilitate this process. The final chapter focuses on project implementation for maintaining and restoring corridors. We describe tools that can help ensure long-term preservation of corridors, including incentive-based conservation and restoration, and offer examples of corridor conservation that are ongoing at multiple scales around the world.

CHAPTER 7

Corridor Design Objectives

Careful corridor design can help avoid the pitfalls discussed in the previous chapter and increase the chance that connectivity goals will be achieved. Though general recommendations about design are always difficult because of the immense variety and even idiosyncratic nature of situations where corridors are implemented, recent research reveals some important considerations of general applicability for establishing and maintaining connectivity. Scale and focal species selection are two important components of design objectives. If focal species are identified, their habitat, dispersal, behavior, and physical needs should be considered. In general, habitat quality, continuity, and dimensions of corridors, as well as landscape context, will affect overall corridor utility.

Important variables that should be explicitly defined in corridor planning are the spatial and temporal scales and the species or communities that the corridor must address. Landscape context, target species, social context, and institutional missions will all influence the project goals and scale. In earlier chapters we discussed spatial scale, in which coarse and fine distinguish differences in grain and spatial resolution. For example, where the goal is for species to be able to move continuously across their entire distribution, connectivity is addressed at a coarse scale. Where the goal is to enable a species to move between habitat patches on a daily basis, a specific corridor may be designed at a finer scale.

Temporal scale, which may refer to the time of year the corridor needs to function or the time period over which it should facilitate species movement, also should be considered. Corridors may be focused on seasonal movements such as migrations, and the time such corridors need to be functional may also affect design. In the case of a seasonal corridor, there may be human activities within the corridor that might be incompatible with the species during migration but are otherwise acceptable. Harris and Scheck (1991) suggested that the time period in which a corridor should function will affect the dimensions required. Some corridors are supposed to function for decades or even centuries and serve entire ecological communities. Harris and Scheck speculated that such corridors should be more than a kilometer (.62 mile) wide. To function for hundreds of years, and particularly if the corridor is important for metapopulation (many populations linked by movements of individuals) persistence, corridors should be at least 100 to 1,000 meters (328 feet to .62 mile) wide. If a corridor is necessary only during a temporary human disturbance of less than one year, then the researchers suggest that 1 to 10 meters (about 3 to 33 feet) in width might be sufficient. The inference from these examples is that specifying the spatial and temporal scales will influence the planning objectives.

Another factor that will influence corridor conservation plans is the designation of focal species. The premise is that corridors should incorporate, where possible, attributes that might enhance use by the focal species. We define *focal species*, also sometimes referred to as surrogate species, as those that warrant special protection; they may be keystone, umbrella, flagship, indicator, specialist, or vulnerable species (Lambeck 1997, Caro and O'Doherty 1999). *Keystone* species are those whose impact on the landscape is disproportionately large relative to their abundance (Power et al. 1996). Wolves (*Canis lupus*) are an example of a keystone species. Lack of wolves led to heavy ungulate browsing and willow reduction in the Greater Yellowstone Ecosystem (Figure 1.5), which has subsequently begun regenerating with the return of wolves to the system (Ripple and Beschta 2004). *Umbrella* species are those whose needs overlap with other species such that their conservation should result in conservation of the other species (Fleishman et al. 2001), while *flagship* species have public appeal for social or economic reasons. *Indicator* species are those whose status is used as a proxy measure of ecosystem conditions (Hilty and Merenlender 2000b). *Specialist* species may be limited by available habitat or other resources, and *vulnerable* species are those listed as endangered or threatened by governments or other groups.

A number of systematic approaches have been proposed for selecting suites of focal taxa, such as Beazley and Cardinal's (2004) efforts in Nova Scotia and Maine and Coppolillo et al.'s (2004) criteria for selecting suites of species for site-based conservation. Because a whole body of literature exists on designating appropriate focal species, we will not delve into selection in detail here but instead highlight two considerations. First, species that are chosen should be easily monitored and identified. Second, selecting a multispecies assemblage is a better approach, because evidence indicates that a single species will not generally represent other species adequately (Hilty and Merenlender 2000b, Roberge and Angelstam 2004). Even where the focus of a corridor is on conserving a single species, we encourage connectivity planning efforts to consider entire communities because the integrity of a community may ultimately affect individual species (Lidicker and Koenig 1996).

One example of designing protected areas and corridors to serve entire communities is California's Natural Community Conservation Planning (NCCP) program. Started in 1991, the program seeks to conserve natural communities at a regional scale while facilitating additional land uses favorable for maintaining the long-term viability of native species. In reality, these plans are usually not proactive but represent a last-ditch effort to address endangered species crises and rapid urban development. They identify multiple species to include in multiple species conservation programs (MSCP) and try to protect their associated habitat types. For example, San Diego MSCP worked with the United States Fish and Wildlife Service, California Fish and Game, private landowners, and other stakeholders to develop quantitative conservation targets for conserving various plant communities. They identified eighty-five focal species, key land parcels needed to conserve those communities, and potential connectivity zones (Figure 7.1). In reviewing the process, the South Coast Missing Linkage project concluded that the NCCP approach lacked a focus on maintaining large-scale processes. In turn, the Missing Linkage project developed an approach that used science-based collaboration to identify fifteen key linkages that would serve to retain connectivity among core areas in the entire South Coast area (Beier et al. 2006). This is an example of conservation projects working at different scales toward similar goals and demonstrates that conservation successes can be synergistic if efforts are coordinated.

Focal Species Considerations

The identified individual species or set of focal species are generally chosen because their persistence is threatened by decreasing landscape

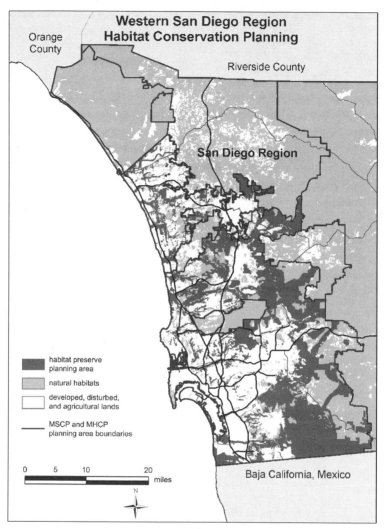

Figure 7.1. A habitat planning area identifying key habitat and corridors to conserve multiple species for the San Diego region as part of California's Natural Community Conservation Planning Program. This conservation plan was cooperatively designed by the participating jurisdictions and special districts in the planning area, in partnership with the U.S. Fish and Wildlife Service and the California Department of Fish and Game, property owners, and representatives of the development industry and environmental groups. The map delineates the Multiple Species Conservation Plan (MSCP) and Multiple Habitat Conservation Plan (MHCP) boundaries. (Permission to use by Susan Carnevale; map provided by the San Diego Association of Governments [SANDAG]).

connectivity. Each species' life history determines its needs and hence affects the variables that enhance or deter its use of corridors. One consideration is the types of habitat that species use on a daily or seasonal basis or during successive life history stages. For many species, the vertical structure of the habitat within a corridor is just as important as the length and width dimensions (discussed later). A second aspect to take into account is species' requirements for dispersal. Another consideration is whether focal species are specialists or generalists. Species that are often referred to as specialists may be particularly problematic because of their restricted habitat requirements compared to more generalist species. Fourth, sociality and other behaviors result in interactions within and among species that alter the distribution of species and can affect the likelihood that focal species will use a corridor. Finally, it is important to be aware of focal species' sensitivity to human activity and of physical limitations that may lead to the focal species avoiding a particular corridor arrangement.

Habitat Requirements

The quality of habitat within an ecological corridor can determine whether the species of interest will use the corridor. Some species will not use a supposed ecological corridor because of low habitat quality (D. A. Norton et al. 1995, Hilty 2001). Native vegetation can enhance the probability that a corridor will be used, while exotic vegetation could deter biota (Bennett 1991). Nonnative riparian vegetation can sometimes dominate a community and exclude native flora and fauna, as in the case of *Arundo donax* (Boose and Holt 1999) and *Tamarix chinensis* (J. C. Stromberg 1997). In other cases, however, nonnative vegetation can serve a structural role or provide food.

Species that evolved in relatively continuous as opposed to naturally fragmented patches are less likely to be able to adapt to human-induced fragmentation and may require more natural corridors. This is especially true where vegetative structure changes dramatically because of humans, whether by the loss or the addition of structural components. A study in South Carolina sprayed fluorescent powder on berries eaten by birds to learn how birds moved and dispersed seeds in an experimentally designed corridor-patch forest and found that birds followed the forested corridors more often than they flew across clearings (Levey et al. 2005). Arboreal species are extremely sensitive to the level of fragmentation, corridor quality, and spatial isolation. In one study in Australia, arboreal marsupials declined by 97 percent in both tropical forest fragments and corridors

(Laurance 1995). In such a scenario, maintaining or restoring corridors may suffice only if they encompass intact forest characteristics.

Whereas some species require access to large areas of relatively homogenous habitat, other species need connectivity among different types of habitats at various life history stages. Many ungulate species move back and forth each year from montane summer habitat to valley winter habitats, sometimes passing through naturally constricted areas. Blockage of such passageways could eliminate a population from a region. Similarly, the survival of some wetland species is imperiled where efforts focus solely on wetland conservation with little or no attention to upland buffer zones or corridors. Though a species may spend most of its life within a wetland, retaining upland connectivity is important for species that require upland habitat for a portion of their life cycle or for dispersal (Semlitsch and Bodie 2003). One such species is the bog turtle (*Glyptemys muhlenbergii*), which has become endangered not only because of wetland loss but also by loss of surrounding upland habitat corridors needed for the turtles to move among wetlands (Klemens 2000).

In some cases, species may require specific natural elements for their survival, and it may be necessary to incorporate those elements into corridor design. They include features such as snags, denning and hibernation sites, and salt licks, as well as other features. For instance, to retain a continuous population of woodpeckers, snags or standing dead trees should be included in the corridor because they are an important resource for woodpeckers (Farris et al. 2004). Similarly, sugar gliders (*Petaurus norfolcensis*), a threatened arboreal marsupial, rely on large old trees for both food and breeding resources and are correspondingly found only where such trees persist, including in linear habitats along streams and roads (van der Ree and Bennett 2003).

Given the potential importance of habitat type and quality, corridor design should incorporate intact natural habitat when possible. Specific requirements of focal species should be reviewed to ensure that those species will continue to have access to needed habitat types over time. If known, special elements required by focal species for survival should be retained in the core habitat areas and corridors likely needed by those species. If the corridor is large relative to species movements, and the species can become resident within the corridor, it will be particularly important to incorporate those habitat needs within the corridor and not just in the habitat patches being connected. In some cases, active management of both the corridor and the surrounding matrix may be required to retain appropriate habitat within a corridor in the long term.

Dispersal Considerations

Understanding the periodic dispersal of a species may also inform the temporal scale that should be addressed so that a population does not become isolated and suffer from associated impacts. Depending on one individual to successfully move from core habitat A to core habitat B every fifty or so years based on the results of a model may make planning for implementation difficult. Also, during the fifty-year period, much change could occur across the landscape, such as the mix of human activities, which might further inhibit animal movement. Moreover, little is known about dispersal patterns for most species, including the range of distances traveled, the success of dispersers, and the likelihood of dispersers using different types of human-occupied landscapes (Chapter 4). Northern spotted owls (*Strix occidentalis caurina*), for example, disperse in random directions, so discrete corridors might not be beneficial for them (Simberloff et al. 1992). It is for these reasons that planning for minimum dispersal should not be the one and only design criterion. A more conservative approach would be to ensure connectivity over shorter time periods.

Like that of animals, plant dispersal strategy can influence successful movement through corridors. A computer simulation model examining seed flow demonstrated that plant dispersal ability influenced the number of species that may occur in riparian corridors across a fragmented landscape (Hanson et al. 1990). The model also suggested that environmental stress, such as flooding, could amplify the problem. In addition, species with higher dispersal capabilities are often aggressive weeds that can displace native species (Kubeš 1996), making it more challenging for native species in the long term.

In some cases, successful plant dispersal may depend on mutualistic relationships. For example, frugivorous birds, especially the mistletoe bird (*Dicaeum hirundinaceum*), disperse mistletoe seeds (*Amyema miquelii*) in Western Australia. As mistletoe birds prefer large habitat fragments, seeds are not dispersed into small remaining habitat fragments (D. A. Norton et al. 1995). Similarly, flying foxes (*Pteropus* spp.) in Samoa play an important role in dispersing and pollinating plant species, and hence their survival and behavior in fragmenting systems will affect the overall forest structure (Cox et al. 1991, Cox and Elmquist 2000). In cases where mutualistic relationships influence dispersal, it will be important to identify and consider both the focal species and the need of the species it depends on for dispersal when planning for connected landscapes.

For these reasons, corridor design should consider dispersal of focal species. Because it is more than likely that little information will exist on focal species' dispersal requirements, corridors should be designed not only to facilitate dispersal. Given the relatively little information currently known about dispersal, it is best to err on the side of caution when designing corridors with the hopes of facilitating dispersal by providing as much interior and native habitat as possible. An evaluation of potential competition with other species also may be important in designing successful corridors. If one species is dependent on another species for dispersal, corridors need to be designed to accommodate the needs of both species.

Generalist versus Specialist

Poorly designed corridors will act as filters. Generalists, such as those that use multiple habitat types or have relatively broad diets, will pass through, while specialists will be impeded (Laurance 1995; Chapter 6). One reason for this is that human-induced fragmentation tends to lead to an increase in weedy species and habitat generalists, whereas rare and sensitive forest-interior species decrease in numbers (Brosset et al. 1996, Dijak and Thompson 2000). Generalist species will dominate narrow corridors, and invasive exotics may also be present (Forman 1995). It is specialists and those species prone to human-wildlife conflicts that most need effective corridors to ensure species survival in a fragmenting landscape (Kozakiewicz and Szacki 1995). Facilitating movement of generalist species such as gray fox (*Urocyon cinereoargenteus*) and red fox (*Vulpes vulpes*), raccoon (*Procyon lotor*), coyotes (*Canis latrans*), and opossums (*Didelphis virginiana*) is easier than conservation of large, wide-ranging, or human-incompatible species such as grizzly bears (*Ursus arctos*), black bears (*Ursus americanus*), and mountain lions (*Puma concolor*) (Harris et al. 1996).

One reason it is more difficult to design corridors for specialist species is that most corridors are not wide enough to retain the specific habitat needs of specialists and are likely to be affected by a host of edge effects (e.g., Hilty and Merenlender 2004; Chapter 5). The edge effects may preclude the availability of specific habitat elements or food items, especially if those resources are dependent on interior, undisturbed conditions. For instance, Mills (1996) suggested that red-backed voles (*Clethrionomys californicus*) may inhabit primarily interior habitat rather than edge or clear-cut areas because the interior forest provides cool, moist conditions that enhance the growth of fungi, a preferred food item.

What we find is that specialists are likely to need wider corridors or larger stepping stone patches with more intact and native habitat in order to maintain the habitat conditions that favor their survival (Selman 1993, Forman 1995, Spackman and Hughes 1995, Kubeš 1996, Perault and Lomolino 2000). For specialists with specific habitat or food limitations, the design of corridors may need to ensure that those elements will be incorporated and retained.

Sociality and Behavioral Factors

Intraspecific and interspecific interactions are important factors to consider in connectivity planning. Social species might not use corridors unless they can move in groups (Laurance 1990; Chapter 6). Likewise, mutualistic relationships, antagonistic relationships, and density-dependent survivorship may influence corridor use (Lidicker and Koenig 1996). We mentioned an example of a mutualistic relationship: the spread of mistletoe being dependent on frugivorous bird behavior. Antagonistic relationships are those in which one species interferes with another species, which may happen when species from the matrix invade a corridor. Brown-headed cowbirds (*Molothrus ater*), a species often associated with edges, parasitize nests of songbirds, decreasing their reproductive success in small fragments and narrow corridors (Hansen et al. 2002). Density dependence is illustrated by root voles (*Microtus oeconomus*), whose dispersal was determined to involve movement toward patches with lower densities, indicating the importance of spatiotemporal demographic variability (Andreassen and Ims 2001).

Another type of variable to consider is behavioral factors that may limit a species' movement. A species may be physically capable of living in and moving across human-impacted regions through corridors or even the matrix but may avoid edges and fail to travel across any sort of disturbed habitat because of behavioral constraints (Lidicker 1999, Lidicker and Koenig 1996, St. Clair et al. 1998). Surprising though it may seem, some birds will not fly across even narrow water gaps or forest clearings (Diamond 1972, 1973, Noss 1991, Machtans et al. 1996), and some large and wide-ranging mammals like mountain lions avoid crossing deforested areas (Smallwood 1994, Opdam et al. 1995). Similarly, American and European species of martins (*Martes*) generally avoid open areas and are considered to be forest dependent (Bissonette and Broekhuizen 1995).

California red-backed voles also appear to avoid cleared areas, perhaps because their primary source of food, truffles, is found generally within forests and rarely in cleared areas (Mills 1996). In an experimental study of butterflies in Natal, South Africa, stands of exotic plain trees and mowed grass had a highly negative influence on butterfly flight paths, causing butterflies to switch directions (Wood and Samways 1991).

Behavioral preferences, if known, should inform corridor design, and that should ultimately improve the chances for successful species conservation. A general assessment of how focal species currently move across the landscape is useful. It is also a good idea to identify and prioritize the conservation of corridors that already serve as natural conduits for movement of focal species (Ndubisi et al. 1995). In some cases, generalist and exotic species may limit use of corridors by specialist species, so that regular control of such species may be necessary. An option requiring less intensive land management would be to retain wide enough corridors that generalist and exotic species, which generally are edge species, do not penetrate the entire corridor.

Sensitivity to Human Activity

Some species do not accept close proximity to humanized landscapes, either because of their intolerance of human activities or because of lack of human tolerance toward them. Noise; air, ground, and water pollution; and lighting can affect whether focal species use a corridor (van Bohemen 2002). One species whose movements have been observed to change because of lights is the mountain lion. A study in Southern California showed that the species avoided corridor sections illuminated by artificial lights and instead used densely vegetated corridors (Beier 1995). Nesting bald eagles (*Haliaeetus leucocephalus*) are sensitive to human activity and will flush from perches, generating higher energy needs. It may be important to regulate human activities along rivers that are priority areas for bald eagles (Steidl and Anthony 1996).

Many top carnivore species that are generalists could theoretically survive and fulfill their energetic needs in heavily human-impacted areas, but such adaptation requires human tolerance of the species. Wolves in the Great Lakes region seem to survive relatively well in a region dominated by agriculture and ranching on private land, despite some conflict with humans (Treves et al. 2004). But wolves are not well tolerated on private lands and multiple-use public lands surrounding Yellowstone National Park. There, at least one individual of every pack known to range outside of the park in the first six years after reintroduction was killed by humans, mostly through legal means (K. Kunkel, pers. comm. 2004). The wolf is an example of a wide-ranging, relatively low-density species, which is often among the first species to go extinct where human activity fragments natural systems (e.g., Beier 1993). Yet wolves can survive quite well in human-impacted landscapes if prey are adequate and they are tolerated by humans.

A review of focal species' tolerance to human activities will be important in designing corridors, especially if human activities fall within the corridor. In cases where little information exists about species' sensitivity to humans, it is best to err on the side of caution. The most cautious approach is to limit human activities within corridors and place corridors far away from high-density human areas. If human activities must occur within or near corridors, they should at least be directed farther away from elements of higher biodiversity value, such as riparian zones. Other steps to limit overall human disturbance include limiting noise, light, pet and livestock activity, feeding of wildlife, and degradation of existing vegetation. For example, earth berms can reduce noise and lighting from sources such as roads, and thick vegetation can filter air particles and water runoff from developments. In circumstances where there are human-carnivore (or other wildlife) interference problems, sometimes conflicts can be minimized through such mechanisms as buffer zones, physical barriers such as fences, active protection, or compensation schemes, all tools to consider in designing linkages (Ahern 1995, W. F. Laurance 1995). Corridor design may also require encouraging compatible human behavior adjacent to corridors, such as storing trash and bird feed properly, discouraging feeding of wild animals, and limiting the planting of vegetation such as fruit trees that may encourage some species to wander out of corridors into human-occupied parts of the landscape.

Physical Limitations

Some species have clear physical limitations that are important to be aware of when designing corridors. For instance, fencing that is too high and has strands close together near the ground can impede pronghorn antelope's (*Antilocapra americana*) and other species' movements (Hailey and DeArment 1972; Figure 7.2). Similarly, researchers speculate that turtles may suffer difficulties in round culverts because of their propensity to try to climb the sides and end up flipped upside down. Moreover, corrugated culverts could snare juveniles or high-center adults. Consequently, square

Figure 7.2. The carcass of a mule deer (*Odocoileus hemionius*) that got caught in a wire fence along Highway 87 in Utah. (Photo by David House, 2003.)

culverts without corrugation may be the most prudent type of underpass to facilitate turtle movement under roads (K. Griffin, pers. comm. 2005). Other types of barriers to consider range from roads and cliffs to high mountain passes and large waterways (Kubeš 1996).

All such potential barriers should be identified in the planning process so as to avoid corridors that look good on paper but in fact do not adequately serve species of interest. In some cases, it may be necessary to replace human-created barriers such as fences or poorly designed culverts with wildlife-friendly designs. It is hard to generalize what makes a wildlife-friendly fence, because that can depend on the species. While a plethora of guidelines exist (e.g., Redland Shire Council 2002, Luers 2004), most generally accept the following standards: three strands of smooth wire, with the bottom strand at least 16 inches above the ground, the second strand at 24 inches, the third wire at 32 inches, and a pole on top at 40 inches. The top pole provides an important visual barrier that wildlife can detect and prevents them from getting tangled in a top wire. Generally, the more visible the materials, the better, so wood, recycled plastic, and PVC are better than wire. Consideration of guidelines, regulations, and incentives for managing the corridors in the long-term is important, as these variables influence the long-term success of corridors (Ndubisi et al. 1995).

Corridor Quality: Continuity, Composition, and Dimension

Beyond assessing potential corridors according to focal species' needs, general site characteristics should be assessed, as they can influence the kinds of species that use corridors, as well as the long-term viability of corridors and the communities that depend on them. Corridor continuity, including habitat quality and composition of plant species, should be considered. Dimensions of proposed corridors are also important.

Continuous Corridor

Here we compare the effectiveness of continuous and discontinuous corridors. A good deal of research supports the importance of continuous corridors as opposed to corridors that are bisected by roads or other activities, especially for more sensitive species (Bennett et al. 1994, Forman 1995, Tilman et al. 1997, Brooker et al. 2001). Unfortunately, few of these studies indicate threshold levels of connectivity below which various species or groups of species would be unable to use the corridor.

Generally, if the habitat within a corridor is too fragmented, passage of species through the corridor may effectively be stopped (Tilman et al. 1997). Gaps and barriers, such as roads, should be avoided where possible (Forman 1995, Brooker et al. 2001). What is perceived as a gap varies from species to species. Gaps of greater than 5 meters (16 feet) in structural vegetation may deter the movement of transient chipmunks (*Tamias striatus*) (Bennett et al. 1994). For mountain lions, major highways that lack appropriate underpasses may prevent connectivity, but highway bridges that accommodate watercourses and some natural vegetation may help maintain connectivity (Box 4.2; Beier 1995).

While maintaining a high level of connectivity is desirable, it is not always an option, especially in more developed landscapes. Fortunately, some species appear to be less affected by discontinuities in corridors than others, but again tolerance of gap size is species specific. For tundra voles (*Microtus oeconomus*) and gray-tailed voles (*M. canicaudus*), gaps in corridors of less than 4 meters (13 feet) did not greatly change or increase transit time (Andreassen, Ims, and Steinset 1996, Lidicker 1999, Wolff et al. 1997). Likewise, research on grizzly bears and mountain lions in the Crowsnest

Pass on the border of Alberta and British Columbia indicates that at least some individual bears and mountain lions managed to cross a two-lane highway (J. Weaver and C. Chetkiewicz, pers. comm. 2005). In that case, maintaining adequate corridors for the animals to approach the highway may be part of the long-term strategy. It is also important to maintain and secure adequate road-crossing structures to ensure that populations on both sides of the highway do not become completely severed from one another.

An additional consideration is the nature of the habitat to be connected. This will depend in part on the focal species and their needs. It is important not to assume that connecting like habitat with like habitat is always the primary goal. As mentioned earlier, some biota require only one specific habitat type; others use several different resources or habitat types in various years or times of the year (Forman 1995). For instance, amphibians and large mammals need pathways to water (Kubeš 1996), so ensuring that they have free access to upland habitat and riparian areas is important, especially in the dry season. Similarly, marsh rabbits (*Sylvialagus palustris*) spend their lives in wetlands but need upland forest-corridor habitats for dispersal (Forys and Humphrey 1996). Many species have varying needs in different life-cycle stages that may require special consideration (Ingham and Samways 1996). Failure to recognize that a focal species needs multiple habitat types throughout its life can lead to poor delineation of corridors and potentially the decline of the species.

A more difficult problem to address is that the way humans perceive connectivity can be different from that of other species. Humans often focus on the physical connectivity of plant communities, yet wildlife may use cues in the landscape that are difficult for humans to see. Mammals might focus on scent trails or other more subtle attributes of matrix habitat. Until more is known about the life history and behavior of species of interest, such sensitivities to connectivity may remain undetected (Lidicker 1999).

For some species and groups of species, general guidelines do exist as to the threshold levels of continuity needed to retain connectivity among otherwise disjunct populations. One modeling effort that assessed corridor functionality for plants suggested that if more than 50 percent of the habitat was destroyed, the percolation threshold (the point at which individuals may no longer be able to move through it) would be reached and a corridor would no longer function to facilitate plant species movement (Tilman et al. 1997). Other investigators have focused on how individual species cross nonhabitat to move among remaining fragments of habitat. For example, a study that examined how gliding marsupials moved among woodland patches in an agricultural system found that most of the movement occurred within 75 meters (246 feet) of the larger patches containing the focal species (van der Ree et al. 2003). Generally, evidence from cumulative research indicates that willingness to cross gaps is species specific (e.g., Ricketts et al. 2001).

Given the importance of continuity for species conservation, increasing attention is being directed toward designing structures to bypass bottlenecks, such as road and railway crossings. With more than 6.4 million kilometers (4 million miles) of roads in the United States alone, roads can be a formidable barrier and a source of mortality for wildlife, which is often associated with vehicle damage (E. Murphy 2005, Forman et al. 2003). Roads are perilous for resident species, especially for dispersers and introduced species that are unfamiliar with them. In Chapter 2, we presented figures on numbers of species killed by roads, including a rare study documenting high mortality for dispersing cougars caused by automobiles. The attempted reintroduction of Canada lynx (*Lynx canadensis*) from the Yukon to the Adirondack State Park in New York, which has many more roads, illustrates how roads can significantly impact naive populations—of the thirty-two confirmed deaths in the park, twenty were by cars (Associated Press 1998). Undoubtedly, roads, which caused a majority of the known deaths, contributed to the ultimate failure of the reintroduction. To reduce wildlife-vehicle collisions, species-specific information about crossing and mortality rates and species data on preferred and successful crossing structures, ranging from vegetated overpasses to bridges and culverts, are needed (e.g., Clevenger et al. 2001; Figure 7.3). Fencing may also reduce wildlife road kill. According to research, road and pipeline underpasses have successfully provided travel routes in many locations and for many species ranging from migrating caribou (*Rangifer tarandus*) and elk (*Cervus elaphus*) to grizzly bear and wolves (Putman 1997, Rodriguez et al. 1996, Clevenger and Waltho 2005). In Chapter 9, we provide a case study involving Highway 93 in Montana, where multiple crossing structures are being built to facilitate wildlife movement. Literature pertaining specifically to such topics as the impact of roads on wildlife and the principles of road-crossing structure design has recently been compiled in a book titled *Road Ecology* (Forman et al. 2003).

Unless contrary evidence is available, the conservative approach of maintaining or restoring maximal continuity may be the best strategy for ensuring focal species survival, given the preponderance of evidence that lack of continuity in a corridor impairs its connectivity value. In addition

Figure 7.3. Wildlife overpass across the Trans-Canada Highway, one of two established to increase connectivity for populations of wildlife in Banff National Park, Alberta. (Photo by Heather Dempsey, Parks Canada.)

to addressing continuity within corridors, this approach may include providing two or more corridor connections between habitat patches. The existence of multiple pathways can enhance the likelihood of a species moving from one patch to another (Figure 4.1). Further, redundancy provides a sort of insurance in case a corridor is affected or destroyed by unforeseen events ranging from natural impacts such as fire to human destruction. In some cases, it may be possible to incorporate different types of pathways such as upland and riparian corridors, each of which may be used by different species. Finally, monitoring corridor passage is important to allow for adaptive management when problems arise and to inform other corridor conservation efforts.

Stepping-Stone Connectivity

Functional connectivity for some biota may not require a continuous connection of relatively intact natural habitat but could involve stepping stones of habitat or protected areas that are not physically connected that certain species can use while dispersing or migrating (Forman 1995, A. F. Bennett 1999; Figure 4.1). Species that are adapted to a habitat mosaic can more easily maintain a metapopulation structure because they are already adapted to disperse in fragmented habitat (Szacki and Liro 1991, Forys and Humphrey 1996). If a species has some ability to cross matrix habitat, clusters of small patches may be an alternative to continuous corridors (Forman 1995, A. F. Bennett 1999). Modeling exercises suggest that for species with limited ability to disperse in a fragmented landscape, stepping-stone patches can greatly increase the ability of the species to disperse (Söndgerath and Schröder 2001).

Stepping-stone connectivity might be better than continuous corridors given the life history of some species. Volant species such as Fender's blue butterfly (*Icaricia icarioides fenderi*) and whooping cranes (*Grus americana*) are two species whose life histories make them well adapted to stepping-stone connectivity (Schultz 1995). Frugivorous pigeons in New South Wales eucalypt (*Eucalyptus salmonophloia*) forests are also adapted to stepping-stone connectivity because eucalypt forests were naturally interspersed within rain forests. Currently, remnant patches of rain forest are surrounded by agriculture, but these pigeons appear to be able to move throughout the patchwork using exotic weed patches as stepping stones, possibly because the pigeons evolved in a habitat type that was patchily distributed (Date et al. 1991).

Keeping stepping-stone habitat functional has been an important part of conservation efforts for red knots (*Calidris canutus*), sandpipers that migrate between southern Argentina and northern Canada. The species has been in decline in recent years. A major concern is that one of their important food sources, eggs from horseshoe crabs (*Limulus polyphemus*), has been declining due to increased harvest at Delaware Bay, a key stopover site on their migration. Lack of this food resource means that red knots are unable to gain sufficient weight to travel to their breeding grounds and breed successfully (Morrison et al. 2004). Key to conserving this species in the long term is ensuring that connectivity of its long migration route remains intact, meaning that the stopover or stepping-stone sites and associated resources that the birds need continue to stay in good condition.

Where a community type has evolved in a more continuous array, a stepping-stone mosaic of patches may serve some species of the community but fail to allow for the maintenance of a successful metacommunity. Therefore, planning for maintaining connectivity through a stepping-stone approach should proceed carefully.

Habitat Quality

Earlier we discussed habitat quality with regard to focal species specifically, but ideally a corridor will serve more than just focal species. Corridors, be they continuous or stepping stone, should function better for all native species if they retain a higher degree of community integrity. A computer simulation by Tilman et al. (1997) suggested that habitat in a corridor must be of higher quality than that in larger core habitat patches to be effective. Survival in marginal habitat might be feasible even given scarce resources. Some species may not use low-quality habitat at all, but the survival, reproduction levels, and ability to move through the corridor of those that do may be compromised. Small mammals that use low-quality hedgerow corridors were found to be significantly less successful in reaching the other end than those that used high-quality corridors (Merriam and Lanoue 1990, Bennett et al. 1994, Forman 1995, Hilty 2001). Vegetative structure and composition both influence habitat quality.

Vertical vegetative structure has been shown repeatedly to affect species richness and composition of biota in corridors (Forman 1995, Hilty 2001). Several studies found that for both native plants and animals adapted to structurally complex habitats, an adequate overstory cover influences species use (Fritz and Merriam 1993, Bennett et al. 1994). Studies of white-footed mice (*Peromyscus leucopus*) suggest that the mice prefer to move along fencerows that are more structurally complex (Merriam and Lanoue 1990), and chipmunks also were found to preferentially use fencerows with tall trees and woodland structure and to completely avoid fencerows associated only with grassy vegetation (Bennett et al. 1994). Likewise, in a northeastern North American farmland, ground-layer native woodland plant species needed overstory cover to survive within fencerow corridor habitat (Fritz and Merriam 1993). For open-habitat species such as some butterflies, open-habitat corridors can facilitate movements through forested areas (Haddad and Tewksbury 2005).

Another habitat component that can increase corridor function is native vegetation, while exotic vegetation can deter use (Bennett 1991). Findings from a study in Queensland, Australia, indicated that corridors that were floristically diverse served arboreal species better than those that were primarily *Acacia*-dominated regrowth (Laurance and Laurance 1999). Nonnative vegetation can, in some cases, replace structures or functions on which the species of interest are dependent such that focal species can use corridors dominated by nonnative vegetation. In Queensland, rain forest birds used forested patches along a creek corridor within which lantana (*Lantana camara*), an exotic species, formed a dense shrub layer. Lantana provided another layer of structure, and the plant's flowers and fruit are potentially useful to birds. In that case, lantana appeared to contribute to desirable structure and serve as an alternative food source, so the corridor might function well for some birds despite the presence of common exotic species (Crome et al. 1994). Where plant restoration is required, however, native vegetation should be encouraged in order to maintain and increase biodiversity.

Low-quality or nonnative vegetation can sometimes be better than no corridor, depending on the focal species. Low-quality corridors enhanced populations of white-footed mice (Henein and Merriam 1990), and dispersing mountain lions in Southern California used a corridor with degraded vegetation (Beier 1995). However, fragmentation-sensitive birds in Southern California did not use revegetated highway corridors as much as remnant strips of native habitat (Bolger et al. 2001). Where connectivity of natural vegetation has been lost, planted corridors might serve at least some species (Bolger et al. 2001).

Designated corridors ideally should be composed of relatively intact native vegetation. To retain native vegetation over the long term, wider corridors that retain less edge habitat are ideal. In some locations, control of nonnative species may be necessary to retain native species. Where habitat is too disturbed to retain native vegetation, simulating the characteristics of native vegetation such as its structure may be useful for some species, but that should be evaluated carefully.

Corridor Dimensions

In addition to continuity and habitat quality, studies indicate that the dimensions of corridors play an important role in determining what species occur within the corridor and the potential speed with which those species pass through the corridor. Here we examine what we know to date about the role of the overall dimensions of a corridor. Both length and width of corridors appear to play a role in the utility of landscape structures to facilitate movement. Vegetative structure could be considered a third dimension, but we discussed that earlier as a component of habitat quality, so we will not revisit it here.

Length is an important consideration. In general, shorter corridors are more likely to provide increased connectivity than longer corridors. Corridors that are too long might not contain some species due to increased distance from core habitat. A study of hedgerows in the Czech Republic concluded that three quarters of forty-one forest plant species examined were not found beyond 200 meters (656 feet) away from the source woods, while the remainder of the species were detected as far as 250 to 475 meters (820 to 1,558 feet) away (Forman 1995). Similar to the findings of the plant hedgerow study, use of corridors along roads by mammals in southeastern Australia decreased farther away from core habitats.

Width is another determining factor in the number and composition of species found in corridors. There has been some debate in the literature about the benefits of narrow versus wide corridors. Scientists hypothesize that narrower corridors may increase the speed with which an animal moves through a corridor (Soulé and Gilpin 1991, Andreassen, Ims, and Steinset 1996). However, the preponderance of data indicates that wider corridors are generally more effective for maintaining connectivity.

Wider corridors generally support more species and thus better uphold community integrity. For example, in the oak woodlands of northern California, fewer native carnivore and bird species were detected in narrower corridors, where activity levels of nonnative species increased (Hilty 2001, Hilty and Merenlender 2004). Wider corridors in Vermont also contained greater numbers of local bird species, and riparian corridors needed to be at least 150 meters (492 feet) beyond the river's edge to incorporate 90 percent of the bird species found in the community (Spackman and Hughes 1995). Similarly, bird richness in Western Australia and along the Altamaha River in Georgia was related to corridor width as well as vegetation degradation (Saunders and de Rebeira 1991, Hodges and Krementz 1996). Width also seems to be a good predictor of bird species richness, nest density, and breeding success in windbreaks (Forman 1995). In Washington state, corridor width was among the variables that influenced faunal diversity (Perault and Lomolino 2000). In Alberta, Canada, forest-dependent birds declined in forest corridors through clear-cuts less than 100 meters (328 feet) in width (Hannon et al. 2002), as did endemic understory birds in Chile (Sieving et al. 2000).

As with fauna, use of corridors by native plants appears to be affected by width. The results of a spatial simulator used to explore the effect of dimensions suggest that width has a much greater effect than length on the probability that plant propagules will be successful, although length remains a significant variable (Tilman et al. 1997). If the corridor is too narrow, plants, which often move by sending undirected propagules, may lose them to nonhabitat or experience edge effects, further contributing to loss of propagules and limiting corridor effectiveness (Tilman et al. 1997). Also, wider corridors are more likely to maintain microclimate variables desired by forest-interior plants (Forman 1991). Researchers in the Czech Republic concluded that species dependent on forest interior conditions such as many forest herbs, fungi, insects, and mollusks could not migrate through corridors 40 meters (131 feet) wide (Kubeš 1996). Likewise, surveys of river corridors in Vermont indicated that widths of at least 30 meters (98 feet) beyond the stream were necessary to encompass at least 90 percent of the riparian vascular plant species (Spackman and Hughes 1995). These studies clearly indicate that narrow strips of habitat may be insufficient to promote connectivity due to the poor quality of habitat and edge effects.

Some researchers suggest that length and width interact such that longer corridors require greater widths (W. F. Laurance 1995). For instance, Beier (1995) suggested that dimensions of at least 100 meters (328 feet) wide and less than 800 meters (2,624 feet) long would be adequate for cougars. However, if the length were to exceed 1 kilometer (.62 mile), he recommended that the width should be greater than 400 meters (1,312 feet) (Beier 1995). One promising, more generalized approach bases minimum viable corridor dimension on the size of remaining patches of habitat (Kubeš 1996). Kubeš estimates that 0.5 to 5 hectares (1.2 to 12.4 acres) habitat patches require corridors not exceeding 1,000 to 2,000 meters (.62 to 1.2 miles) in length and be at least 10 to 20 meters (33 to 66 feet) in width. Larger patches (5 to 50 hectares, or 12.4 to 124 acres) would need corridors of no longer than 400 to 1,000 meters (1,312 feet to .62 mile) long and at least 20 to 50 meters (66 to 164 feet) wide to maintain connectivity.

Further research on corridor dimension is desirable to provide land managers and conservationists with definitive guidelines that may be widely applicable. Lack of clarity contributes to corridor width arguably being one of the most contentious issues, especially across private land, for example, where stream setbacks may be required and result in restrictions on land use (Dybas 2004). Available data do offer some generalities that can be useful in comparing one corridor design to another corridor design. Choosing sites that minimize length and maximize width is ideal for continuous corridors. For species that are sensitive to edge effects, maintaining corridors that are wide enough to retain some habitat not affected by edge is important. The same concept applies to stepping-stone corridors, where each patch should be large enough to maintain the integrity of the patch. Edge effects vary by habitat type, but, as discussed in Chapter 5, the effects can often permeate several hundred meters beyond an edge. As mentioned previously, one of the inherent dangers managers and scientists face is that focal species might perceive corridors differently from our expectations and so not use them as anticipated. For that reason, perhaps we should err on the side of making corridors as large and encompassing as possible and incorporating as many favorable attributes as we can (Ingham and Samways 1996).

Landscape Configuration

In identifying corridors, it is also important to consider how the surrounding matrix may affect the utility of the corridor (e.g., D. A. Norton et al. 1995, Perault and Lomolino 2000). It is clear from an increasing number of studies that species occurrence at any place in the landscape may not only depend on the immediate characteristics at that spot but also be affected by characteristics of the larger landscape (e.g., Bolger et al. 1997, Gascon et al. 1999, Ricketts et al. 2001, Hilty et al. 2005; see also Chapter 5). This means that in assessing and planning connectivity, it is important to consider current and forecasted landscape geometry, since the location with respect to other landscape elements can influence use (e.g., Kubeš 1996, R. Walker and Craighead 1997). For instance, our research conducted in northern California's oak woodland and vineyard matrix examining the distribution of mammalian predators, demonstrates how the configuration of the landscape could influence the current and future probability of the occurrence of predators (Hilty and Merenlender 2004, Hilty et al. 2006). The results of our study indicated that the configuration of vineyard matrix appeared to influence the distribution of mammalian predators. In general, native species had lower probabilities of presence in large vineyard blocks than in isolated vineyards surrounded by natural habitat or within natural habitat. The highest probability of occurrence for nonnative species, in contrast, fell within large blocks of vineyard.

The fact that wildlife use of the matrix can depend on the landscape configuration of such an area is increasingly being appreciated by conservation scientists (Åberg et al. 1995, Dunn 2000, Kremen et al. 2004). One study in California farmlands showed that landscape configuration and proximity to wildlands influenced crop pollination services provided by native bee communities. Pollination services were higher in farmland directly adjacent to natural habitat, suggesting that native bees are using managed lands only on the margins (Kremen et al. 2004). Similarly, wolves in northern Italy were most active in agricultural land adjacent to forested areas (Massolo and Meriggi 1998). Other studies support the idea that wildlife may be more likely to pass through the matrix if individuals can remain closer to habitat patches or continuous intact habitat (Downes et al. 1997, Perault and Lomolino 2000).

The finding that landscape variables may influence species distributions is important for several reasons. A high probability of species occurrence is indicative of conditions where species are likely to have higher densities and thus be more resilient to human and environmental perturbations (Araújo and Williams 2000). Persistence of any species in a region depends on the density and proximity of other populations in the region. Both are likely to change as habitat is converted and fragmented. Locations with low probabilities of occurrence or small isolated populations are more likely to experience species extinctions (Chapter 3).

Other studies also indicate that species occurrence and richness in habitat patches and corridors may be influenced not just by local habitat variables in the patch but also by landscape variables farther from the patch. A study in the southern Appalachian Mountains of eastern North America indicated that relative abundance of songbirds was significantly correlated with landscape variables, including both composition and pattern variables up to 2 kilometers (1.2 miles) from the points surveyed. Local habitat variables were, however, more important than the landscape-level effects (Lichstein et al. 2002). Variation in the impact of landscape variables on songbird numbers could be related to the level of intactness of the surrounding landscape and is probably also species specific (Bolger et al. 1997).

Species may have different scales of response such that some species may be found farther from natural habitat patches than others (Ricketts et al. 2001). Similarly, some species will use fragments of habitat only if the fragments are not beyond a certain threshold distance from core habitat (Åberg et al. 1995, Dunn 2000). Diversity of habitats nearby (Drapeau et al. 2000, Wright and Tanimoto 1998, Hilty et al. 2006) and structure or cover remaining in the human-impacted areas also may affect species presence at a given point in the landscape (Bentley et al. 2000).

These studies illustrate the importance for conservation planning and management of looking beyond habitat loss and fragmentation toward understanding how species relate to different landscape elements. If the overall integrity of natural habitat is low in a given landscape, it may be more difficult to achieve connectivity with either continuous or stepping-stone corridors. If there are options, a good strategy is to place corridors in more intact landscapes. Because it is difficult to maintain and restore biodiversity in a heavily humanized landscape with core areas and corridors, Schmiegelow et al. (2006) propose a paradigm shift in which conservation lands are the supportive matrix and development activities are carefully managed so as not to erode other values. Thus, the islands are where more intensive human activities occur, set within a continuous conservation landscape. This concept has merit especially in biomes that are still relatively intact, such as the boreal forests. However, corridors will continue to be an important conservation tool, probably largely because they are intuitive to a broad audience and are especially relevant in landscapes that already experience a moderate-to-heavy human influence. Also, simple steps can be taken, in some cases, to make it possible for human-altered lands to enhance connectivity. Management of the timing and technique of mowing fields is an example of actions that can minimize disturbance and impact to nesting birds and other species. Mowing from the center of a field toward the edge, for instance, can improve wildlife survival (Natural Resources Conservation District 1999). Matrix management was discussed in more depth in Chapter 5.

First and foremost, a corridor should link core areas of habitat to facilitate more continuous populations, or where that is not possible, to enhance movement among distinct populations (Galle et al. 1995). Landscape elements such as rivers and rock outcroppings that may be barriers to focal species should be incorporated into planning (Kubeš 1996). In addition, abiotic factors, which can deter use, should be considered. For example, wind direction can affect seed and pollen transfer. Similarly, transport of seeds by water should be considered, and migration routes of animals carrying seeds and pollen could be important in determining the location of successful corridors in the landscape (Kubeš 1996).

Many corridors, both de facto and planned, encompass riparian zones, and maintenance of riparian corridors can protect hydrological processes (Harris et al. 1996). In addition, rivers and streams are natural elements in the landscape that guide animal movement (Noss 1991, Ndubisi et al. 1995, R. Walker and Craighead 1997), and sufficiently wide riparian corridors can encompass a gradient of communities that will facilitate movement of many native species (Harris et al. 1996). Various species from black bears to forest-dependent birds have been found to successfully use well-buffered stream zones (Harris et al. 1996, Hannon et al. 2002). Riparian zones tend to have less exposure to wind disturbance, and vegetated streams facilitate the movement of many aquatic and semiaquatic animals, such as otters (*Lontra [=Lutra] canadensis*) and mink (*Mustela vison*), that are sensitive to fragmentation (Harris et al. 1996, W. F. Laurance 1995). In addition, many species may require regular access to water, and stream gullies retain moisture and often contain deep alluvial soils that promote vegetative growth (W. F. Laurance 1995). Furthermore, riparian areas often contain diverse flora and fauna and are generally highly productive (Harris et al. 1996). In arid systems, the areas of highest biodiversity often fall within the riparian zones. The results of a study in Queensland, Australia, for instance, illustrate that riparian corridors harbor more bird species than human-planted windbreaks (Crome et al. 1994), and a study in California indicated that mammalian predators were eleven times more likely to be detected along riparian corridors than in the upland matrix (Hilty and Merenlender 2004).

Some species, however, do not use riparian zones and might need upland corridors. Voles such as *Arborimus longicaudus* and *Clethrionomys californicus* probably would not use riparian corridors, so upland corridors would be needed for them (Simberloff and Cox 1987). Also, riparian zone vegetative composition may vary dramatically from that of other terrestrial communities (Cross et al. 1991), so some species might not be represented within a riparian corridor. Disagreement remains as to the importance of maintaining upland corridors, with some researchers arguing that while upland corridors have an important function, wide riparian corridors that contain upland vegetation should function adequately (Harris et al. 1996). The river continuum concept (Vannote et al. 1980) emphasizes the importance of protecting both habitat types to ensure ecosystem process and function.

Restoring Corridors across the Landscape

Given the importance of connectivity and the challenges of restoration, identification and conservation of corridors before a landscape is fragmented are desirable, rather than trying to reconstruct corridors after fragmentation (Ahern 1995, Kubeš 1996). In many cases, it is too late to plan for connectivity, as fragmentation of natural habitat has already occurred, especially in urbanized areas and areas of high human impact. In such landscapes, it may be necessary to create or restore corridors to maintain species or a community in the region. While many of the factors that should be included in this process are the same as those we have already discussed, we will briefly describe some additional factors and tools that may be useful in efforts to reconnect landscapes. There is a whole field of professionals, landscape architects, who draw up plans for corridor restoration (e.g., Fleury and Brown 1997). We will not cover the field in depth here. Instead, our goal is to explain the major elements in such processes and direct interested readers to other sources for more specifics about site restoration (e.g., Society for Ecological Restoration International Science and Policy Working Group 2004).

In order to achieve restoration, it is necessary to define what restoration is meant to achieve. While restoring a corridor between habitat fragments might seem on the surface to be a good idea, it needs to be clear that the species remaining in those fragments will benefit. If the habitat patches are heavily affected by human activities or nonnative species, it may be that corridor restoration is not the best initial conservation action.

Where it is clear that reconnecting parts of the landscape will enhance long-term conservation goals, the restoration potential should be evaluated. There is no cookbook solution to stream restoration that will work in a wide variety of sites and conditions. The topography and soil may be important physical variables to consider (van Bohemen 2002). If water is important for the success of restoration, it will be important to evaluate the geomorphology and hydrologic flow regimes. Much effort could be expended in planting native species, but those species may not grow in some soils or in soils that have inappropriate water regimes. Likewise, selecting seed sources may be important. Some varieties of a species will do better in a specific location than others. In general, it is desirable to use seed sources of native species, harvested ideally in the local vicinity or at least of the same varietals found in nearby natural habitats. Finally, as discussed, the potential dimensions of the corridor should be considered relative to the focal species that need the corridor.

If the site is deemed appropriate for restoration, another step is to decide whether to use active or passive restoration. The restoration planners should consider the needs of focal vegetation and wildlife, such as growing conditions for plants and hiding places for wildlife (Figure 7.4). Active restoration efforts, such as planting of vegetation, can bring back plant diversity and structure quickly in some settings and are often necessary in highly modified settings (e.g., Sternberg 2003). However, the effort and costs required restrict the utility of those methods, so active restoration is best applied in spatially small areas. Passive restoration, particularly in areas where there is adequate water, has resulted in the natural regeneration of

Figure 7.4. A representation of critical structural attributes that landscape architects should consider when designing corridors for birds in southwestern Ontario. Two or more rows of conifers (fir, pine, larch, spruce, red cedar), one or more species of deciduous trees (apple, beech, elm, oak, willow), multiple rows of shrubs (multifloral rose, sumac, dogwood, hawthorn, berry bushes), grasses or sedges, horizontal limbs, and snags are all recommended (Fleury and Brown 1997). (Reprinted from Fleury and Brown © 1997, with permission from Elsevier.)

large areas of forest and is often much easier and cost effective (Brooks and Merenlender 2001, Floursheim and Mount 2002). Fencing to keep out large herbivores is a tool commonly used in passive riparian restoration that can allow riparian vegetation to fully recover (Figure 9.2; Opperman and Merenlender 2000). In some cases where it is possible to restore a corridor, continued maintenance may be necessary for long-term functioning. Also, other human impacts such as light and noise pollution may have to be mitigated, as discussed earlier.

In many cases, restoration projects are not focused just on restoring connectivity for plants and animals, but also have a human aesthetic or recreation goal (van Bohemen 2002), such as in greenway planning in urban areas. In densely human-populated areas, it may be especially important in restoration projects to provide for both human benefits and wildlife enhancement in order for society to accept restoration plans. In such cases, additional planning is required to minimize conflicts between potential human and wildlife uses. For example, many species avoid areas of high-intensity human recreation (A. R. Taylor and Knight 2003). If the corridor is to include recreation trails, it may need to be much wider to function for focal wildlife species and may require more maintenance to keep focal plant species from being trampled or degraded.

One example of a successful effort to restore a corridor is in Banff National Park in Canada, where human infrastructure and activities in the identified potential corridor limited wolf use. Human activities were greatly restricted in the park through eliminating some unnatural features, including a bison enclosure, a camp, a barn, and horse corrals, and closing an airstrip. While some human activities such as hiking, climbing, and car traffic continue in the valley, wolf activity increased sixfold, an indication that the restoration has been successful for that species (Canadian Parks and Wilderness Society 2005, Duke et al. 2001).

As discussed in Chapter 4, roads are a ubiquitous landscape feature throughout the world and have the potential to sever once connected wildlife populations. Increasingly, research and engineering efforts are developing new methods to mitigate the risk of wildlife deaths caused by roads and to develop mechanisms to retain connectivity despite roads. Researchers in Europe found that one relatively cost-effective way to reduce bird deaths was to build an incline on the side of the road at least 1.5 meters (5 feet) in height to provide birds lift above oncoming cars (van Bohemen 2002). Other species might require underpasses or overpasses. When highways are being upgraded or new roads are being installed, that is a good opportunity to assess potential impacts to wildlife and even vegetation. Extensive research, strategy, and design of road-crossing structures have been conducted in Banff National Park in Canada (e.g., Clevenger and Waltho 2005; Figure 7.3), in the Everglades of Florida (e.g., Foster and Humphrey 1995), in the Netherlands (e.g., van Bohemen 1996), and elsewhere. These efforts provide increasingly specific design specifications for a range of species, from tunnels for amphibians to overpasses for ungulates and bears (Forman et al. 2003).

Where active corridor restoration is a goal, incorporating a landscape architect on the design team will be important. Selecting active or passive techniques must be based on a site evaluation, and potential conflicts between wildlife and humans should be mitigated. Any plan should incorporate monitoring before and after restoration to allow for adaptive management such as addressing unexpected conflicts between wildlife and humans.

In summary, although scientific research on this subject is far from sufficient, land-use planning still needs to proceed (Watchman et al. 2001). Most published corridor research has focused on species least likely to need corridors to persist and, moreover, has been conducted at scales with limited relevance to land-use planning (Barrett et al. 1995, Perault and Lomolino 2000, Bennett et al. 1994). This makes it all the more important to ensure that monitoring and evaluation are part of any corridor implementation plan, so that we can refine our techniques. Also, because most corridors will be affected by surrounding human activities, monitoring is a tool that will help direct long-term management of corridors to ensure that they remain functional.

CHAPTER 8

Identifying, Prioritizing, and Assessing Corridors

The conservation objectives of any project will influence the type of corridors being considered, as well as the most desirable site and landscape-level characteristics. Those characteristics, in turn, will influence site selection for protection or restoration activities. The focus of this chapter is on planning for connectivity primarily through establishing and conserving corridors. We start by reviewing the importance of collaborative conservation for successful corridor planning and implementation. The issue of scale is described in detail because of the influence that scale has on project planning and implementation. Then various methods to help identify, prioritize, and assess specific sites that may meet connectivity objectives are described.

Technological advances in computing power, analysis software, and remote sensing make it possible to view and quantify landscape features more easily than ever before. New modeling approaches are being used to predict where species are likely to be found across the landscape and how they might move through it. Advances in remote monitoring are allowing researchers to track wildlife better than ever before. A combination of methods discussed in this chapter should help systematic planning for corridor conservation and restoration.

Collaborative Conservation Process

We view initiating a collaborative process as the most important first step in corridor planning and the best way to capture essential local knowledge about the project setting. Collaboration provides a mechanism for incorporating the expertise and desires of local experts and community members into the planning process. Because the sociopolitical realities that communities face are important, especially when it comes to project implementation, stakeholders' diverse opinions should be incorporated as well. And if the ultimate decisions will be made by certain boards or politically appointed representatives, involve at least a few of those people throughout the process. If they can voice their thoughts, understand the project goals and constraints, and have confidence in the process, they will be better able to assure other influential decision makers that the best approach is being taken. Engaging conservation scientists in the planning process is also important and will help incorporate the best possible available scientific information, given realistic time and budget constraints, and ensure that the outcomes will be defensible. Input from land managers, wildlife biologists, hydrologists, landscape architects, resource economists, planners, and naturalists is highly desirable.

In an attempt to define a common language for this approach to decision making and to share lessons learned from case studies, the Sonoran Institute published a useful report titled *Beyond the Hundredth Meeting: A Field Guide to Collaborative Conservation on the West's Public Lands* (Cestero 1999). Although focused on public land issues, this report offers helpful guidelines to improving the chances of success of the public processes that are a necessary part of planning most corridor projects. In the report, efforts that are place- or community-based are distinguished from those that address a specific policy or interest-based initiative. The former involves a group of people who are all interested in the same place but do not necessarily come from or live in the area of interest; for example, local watershed groups that have formed in many places may include people who recreate in an area but do not live there. Interest-based initiatives are usually formed by public representatives to assist with policy development, such as development of a regional plan or local ordinance. A good example of policy-based initiatives that require stakeholder participation is habitat conservation plans designed to protect habitat as compensation for future takings of endangered species in the United States. Other environmental regulations can require that habitat loss be mitigated, and conserving corridors is one of the more common actions taken. In fact, recent proposals have been made to amend the California Environmental Quality Act so that impacts to wildlife corridors could be avoided, and mitigating for those impacts would be mandatory (Schlotterbeck 2003).

Cestero (1999) reports that place-based efforts work best when they are led by local participants rather than government representatives and take place in an open and inclusive process that can accommodate a full range of perspectives, including those of government representatives. It is better if participants do not try to represent large interest groups, because confusion can arise when individuals are held accountable for a large, diverse interest group, some of whom will feel their interests were not well represented by those representing them in the process. For example, having several individuals supposedly representing the "environmental community" is less likely to work than having someone represent the interests of local duck hunters. In addition to completing the desired projects, collaborative conservation can lead to an increased capacity of community residents to respond to external and internal stresses that will inevitably arise. That can help communities prevent future problems from becoming crises.

Another way to incorporate stakeholders is to present preliminary ideas to a representative group (i.e., of the public) at discrete times throughout the planning process to incorporate their feedback. This method may be more effective—and may prevent opposing interest groups from derailing the process—when issues that polarize the community are being addressed.

Articulating clear objectives and agreeing on a list of goals or stated desirable outcomes are important first steps for any group process. The goals are entirely context specific and will be influenced by the physical and biological characteristics of the surrounding environment, scale of the project, institutions involved, existing policies, financing, and other factors. There are usually multiple goals for any project, and they should be identified individually even if they overlap. For example, five main aims were articulated for the European Coastal and Marine Ecological Network: (1) raise awareness of the importance of a network approach (protected core habitat areas linked by corridors) to the conservation of habitats and species at a Pan-European level; (2) provide a platform facilitating coordination and cooperation between existing and proposed networks at local, national, and regional scales; (3) identify gaps in the current approaches to site and species conservation; (4) provide the necessary scientific information to inform the process of network development, adding value by setting the Pan-European context for individual initiatives; and (5) provide support for local and national initiatives in network building (European Center for Nature Conservation and Council of Europe 2003).

Project goals often have implicit hypotheses that need to be tested. For example, if corridors are being conserved to facilitate wildlife movement between habitat patches, it is likely that the utility of the designated corridors for species movement remains untested, and it is important to recognize these types of untested assumptions early in the planning process. Having a group begin by collecting and evaluating existing information can be a first step toward testing fundamental assumptions about the system; it will increase their understanding of the system and is one way to empower a group early on (Cestero 1999).

One clear conclusion from those who have studied examples of collaborative conservation is that groups focused on smaller areas are more likely to succeed (Cestero 1999). This is because those involved can relate to the landscape in question, and regular participation from people spread across a large geographic area is not required. The Quincy Library Group in northern California is an example in which a group of approximately thirty people developed a plan for 2.5 million acres of public forestland that in the end did not adequately attend to the diverse interests represented in that large and relatively populated area (Duane 1997). Such larger-scale conservation projects are better addressed through a network of local efforts (Cestero 1999).

Addressing Scale

The spatial and temporal scale of a corridor plan is the most defining aspect of the project, as emphasized throughout this book. The geographic extent of the project must be defined early on in the planning process, even though, in reality, the geographic extent of an ecological network under consideration often changes over time. This may be because of changes in project goals that can result from changes in the institutions and individuals involved in resource management in the area. For example, a single landowner may be required to conserve a wildlife corridor as mitigation for developing the site. At a later time, a watershed plan may be developed that prioritizes habitat linkages for the entire surrounding watershed. This is one reason planners of local corridors should consider the broader landscape context even if there are no existing plans to implement broader ecological networks.

Temporal and spatial scales, as defined in Chapter 2, influence the inferences we make about landscape patterns and processes. The scale at which measurements are taken influences our ability to detect spatial patterns (see Figure 2.2). For example, the perceived number, sizes, and shapes of patches in a landscape are dependent on the linear dimension of the map used (Gardner et al. 1987). Maximizing the representation among habitat types is often a goal for protected area planning and corridor conservation, but the optimal sites to protect to ensure adequate species representation will change depending on the geographic extent of the analysis. Geographic extent also greatly influences the type of data that will be useful and the outcomes of any prioritization effort.

Both biotic and abiotic processes vary in their operating scale (Gosz and Sharpe 1989, Wiens 1985). For example, anadromous fish are affected by an entire river system and ocean environment. In contrast, some minnows are influenced primarily by processes that occur within small, intermittent tributary streams. Likewise, the hydrologic process of a river operates on an expanse that includes multiple watersheds, associated upland processes, and weather patterns; an upland tributary is influenced by a smaller watershed. Perhaps more interesting is the fact that stream flow from the upland tributary is likely to be more variable over a shorter time period, and therefore finer-grain temporal data are required to accurately measure changes in the hydrograph (a graph of stream discharge over time) compared to what would be required along a river's main stem. In this case a coarse-scale measure of a large-scale process is sufficient, while a fine-scale measure of a small-scale process is required to gain a thorough understanding of the system.

Generally, we sacrifice grain when we increase extent of a particular study system. It is often impractical to measure large-scale systems at a fine-grain resolution. Ecosystem variables are often measured at a smaller scale in the field, and then the information is scaled up to try to address the larger scale. The opposite is often true for geographic data, which are easier to obtain digitally at a coarse resolution for large geographic areas and may need to be applied to smaller areas, resulting in errors due to insufficient resolution of the original data. This is an important point because much of the geographic data used for planning today are collected digitally.

Digital geographic data come in a variety of accuracies and scales and from a variety of sources. It is important to differentiate between local, regional, and global remote imagery and geographic digital data. Local or small-area data are exemplified by airborne photography at a resolution from 1:200 to 1:10,000; regional or middle-scale data can be found in aerial photography or satellite images from 1:10,000 to 1:80,000 resolution; and global scale is exemplified by satellite images from 1:80,000 to 1:1,000,000 resolution (Gyorgy 1994). Usually, the larger the geographic extent of these data, the lower the resolution or grain that can be achieved, leading to increased chance for error. However, the advantages of coarse-scale data can sometimes be lower costs and easier data storage and computation.

It is also essential to understand that the minimum mapping unit of any spatial data coverage may be selected to meet specific project objectives. An example of this is the California Gap Analysis program, which maps land ownership and habitat types using a minimum mapping unit of 200 hectares (494 acres) for uplands and 80 hectares (198 acres) for wetlands. This may be appropriate for state and regional landscape analysis but is far too large for more local analyses (Beardsley and Stoms 1993).

Research and application objectives set the working scale of a project, which in turn influence the type of digital data that will be useful. The scale of a project designed to address connectivity influences the types of analysis that may be appropriate for site identification, assessment, and prioritization. For example, some efforts to reconnect fragmented landscapes are focused on a regional scale, such as within the Greater Yellowstone Ecosystem (Hansen et al. 2002) or across entire countries such as Australia's WildCountry project, which focuses on national-scale landscape connections (T. McDonald 2004). Other efforts are focused on individual watersheds, like the Mountains to Mangroves project in southeast Queensland, Australia (www.mountainstomangroves.org/), or even on single barriers to movement such as a highway (W. McDonald and St. Clair 2004).

State and national priorities are usually focused on monitoring natural resources and the goods and services they provide to the entire public, and therefore low-resolution data for large geographic areas are collected for that purpose. Clearly, there is a role for data at all scales; however, freely available digital data are often too coarse for local conservation planning projects. Problems often arise when coarse-level data are used for site-level decision making. Extrapolation, the process of estimating information from a given data set, can be problematic when we try to extrapolate from data with a given grain and extent to another area with different dimensions requiring scale transformations. There is a threshold beyond which scale transformation is not effective. For example, classified satellite data often have a limited resolution of 30 by 30 meters (98 by 98 feet) and therefore are not informative at a finer scale. However, the limitations of coarse-scale data are being recognized, and more local governments are investing in higher-resolution data for local planning efforts. Those data can include cadastral (land ownership parcels) data and digital aerial photographs that are mapped to coordinates on the ground and corrected to match the underlying terrain with a resolution of 2.5 meters (8 feet).

A Geographic Information System (GIS) can help integrate physical and biological information relevant to corridor identification, assessment, and prioritization, but one must pay attention to the grain and resolution of the data used and the scale of the project objectives. Despite the limitations, it is better to have coarse information for corridor conservation planning than no information.

Identifying Corridors for Conservation and Restoration

Corridor site selection is critical and greatly influences the success or failure of any attempt to conserve and enhance connectivity. Relevant methods of site selection have evolved in part from research on systematic conservation planning. In their seminal paper, Margules and Pressey (2000) list the following stages for protected area planning, which are not unidirectional; rather, each one will influence the others:

1. Compile data on the biodiversity of the planning region.
2. Identify conservation goals for the planning region.
3. Review existing conservation areas.
4. Select additional conservation areas.
5. Implement conservation actions.
6. Maintain the required values of conservation areas.

The same type of framework was described for ecoregional planning (Groves et al. 2002, Groves 2003). These principles can be used to identify and prioritize habitat corridors. In fact, reserve planning and corridor planning are often one and the same for very large-scale linkage efforts that may operate on a national scale. This is because the goal of such efforts is often to identify new reserves that will increase connectivity among existing reserves.

Corridor site selection is generally focused on increasing connectivity between remnant habitat to facilitate animal and plant movement. A useful general framework was developed to provide landscape architects with a process to incorporate habitat quality in the design of corridors in rural-urban fringe areas (Fleury and Brown 1997). This framework provides very coarse steps that should be included in the planning process, such as identifying the project goals, desirable corridor characteristics, and focal species, in addition to many useful design features. A more specific check-list, developed to guide planning efforts focused on regional networks (Foreman 1999), attempts to integrate the protection of special elements, representation of environmental variation, and conservation of focal species (Noss 2003). The methods described in Noss's checklist include most of the elements of systematic conservation planning, along with most of the modeling methods employed to address species conservation. Many of the methods listed depend on extensive amounts of data and can re-quire complex analysis. That is why working with scientists to ensure rig-orous analysis of special elements, representation, and focal-species requirements is emphasized by Noss.

It is easy for conservation scientists to draft a laundry list of data, analy-sis, and models that would improve systematic conservation planning, but focusing on essential steps that can be accomplished on a fixed budget may be more helpful to practitioners. So, while we will include an explana-tion and examples from many of Noss's recommended methodologies, we also highlight steps that are easier to accomplish, based on our real-world experience of working with local conservation organizations on fixed budgets (Box 8.1). We know that those trying to conserve land in specific circumstances do not want to spend large amounts of time or resources modeling a system that requires more immediate action; however, in many instances, models can provide insights not possible from evaluating scant data. Box 8.2 includes general steps that we suggest should be addressed in systematic corridor planning efforts, depending on the objectives and scale of the project. Many of these are also found in the cited references and general conservation planning texts (Groves 2003).

Different analysis strategies may be required, depending on the extent to which the landscape has been converted and the level of fragmenta-tion of native vegetation. There is a continuum from entirely altered to relatively pristine landscapes. In highly disturbed systems, small patches of remnant vegetation may persist but often require intensive management to curtail impacts from invasive species and other disturbances. Here and in highly urbanized landscapes, linkages have to be created by either pas-sive or active restoration. *Passive restoration* refers to areas that will regen-erate on their own if protected from disturbances. An example of passive restoration is fencing creeks from grazing and browsing and allowing plants in situ to regenerate (Opperman and Merenlender 2000). Other sites may require active restoration. For example, dryer hill slopes of Cal-ifornia's North Coast are entirely dominated by European annual grasses

Box 8.1.

Sonoma County Case Study of Prioritization (Newburn et al. 2005).

In 1999, Adina Merenlender initiated a collaboration with the Sonoma County Agricultural Preservation and Open Space District to develop a plan for open-space acquisition in Sonoma County, California. The District was created by voters in November 1990 and funded with a local sales tax. We worked with District planners and decision boards to identify and prioritize desirable conservation benefits. Initially, we compiled available digital data on open-space and agricultural and natural resources of Sonoma County from available databases. The District authority defined four categories of benefits: agriculture, greenbelts, natural resources, and recreation. The available mapped information related to those topics was reviewed, critiqued, and updated by a broad spectrum of local agricultural, planning, and natural resource professionals. It quickly became clear that the benefits categories were primarily nonoverlapping in space and that each benefit type had distinct supporters; therefore, separate funds were established to address each benefit type individually.

We developed parcel-level models of land-use change and easement values using a digital parcel map for Sonoma County (many other local governments are now in the process of getting their parcel boundaries into digital format). The models allow staff planners to make tradeoffs for the relative development threat and acquisition cost for each available parcel. Finally, we helped county computer programmers design a user-friendly GIS interface that allows district staff to query any number of parcels and generate a report of all the mapped benefits included in each site.

Our approach to date in making the District's interactive GIS planning tool useful for the planners is to provide them with information on benefit types, acquisition cost, and relative threat for each parcel and allow them to weigh these in setting priorities. We recognize that conservation planners can best integrate this information and adjust the analysis required depending on changing demands from the District and partnering institutions and other political realities. For example, the District staff used the information to solve diverse problems, from identifying the best places to implement a fast-track easement program for riparian corridor setbacks to selecting upland sites that meet California's criteria for allocation of bond funds for oak woodland conservation. The District's acquisition priority areas create a landscape-level plan, while still providing the flexibility to respond to unexpected conservation opportunities. This flexible interactive acquisition plan has eased public concern over subjective decision making, allowed transparent exploration of alternative conservation strategies, and reduced the time it takes for project approval in what was formerly a heavily politicized process.

These types of collaborative efforts take time that most conservation biologists do not have and some resources that many smaller land trusts do not have. However, conservation biologists are often in a position to influence conservation organizations or, in some cases, to work for those organizations. Based on this experience, we offer the following practical suggestions for land conservation institutions:

1. Initiate an early collaboration with conservation scientists and work toward explicit conservation objectives. Then examine the distributional relationships (e.g., correlation and relative variation) among biological benefits to determine whether benefit types should be targeted jointly or individually. If benefit distributions rarely overlap or the relative weights between benefit types are difficult to quantify, consider allocating the budget into separate funds for each conservation objective.
2. Allocate effort for biological data collection and to increase the spatial resolution of land-use and land-cover information to better address the conservation objectives and landscape context. If a particular component (benefit, cost, threat) exhibits high spatial variability, greater effort should be allocated to that component to reduce uncertainty in predictive modeling.
3. Model scenarios of land-use change and assess the expected loss of benefits for each conservation objective across the landscape.
4. Estimate the relative economic costs for conservation easements or fee title acquisition across the landscape.
5. Develop a simplified user interface that overlays model results for available land parcels. This allows the user to weigh that information and make informed tradeoffs depending on purchasing options, funding partners, and political realities.

Box 8.2.

Steps to Identify, Prioritize, and Assess the Location of Corridors

These are not unidirectional but rather represent important components of the planning process that will influence one another.

Prior to identification:
• Assemble a planning team.
• State purpose of corridor project.
• Define the geographic extent of the project.
• Determine if existing remnant habitat is sufficient or if restoration will be required.

Identify potential networks:
• Acquire relevant data from digital sources, experts, and publications.
• Integrate data into a GIS.
• Perform land mapping.
• Identify developed areas and barriers to individual movement.
• Map available information on special features and focal species.

Prioritization:
• Map potential corridors.
• Estimate cost.
• Perform vulnerability analysis.
• Weigh benefits, costs, and threats for various site options.

Assessment:
• Evaluate site characteristics and quality.
• Survey for animals and plants.
• Identify potential limiting factors.
• Solicit feedback from the community.

where blue oaks (*Quercus douglasii*) were removed for range improvement. Active tree planting and irrigation are necessary to restore blue oak woodlands (Brooks and Merenlender 2001). *Active habitat restoration* often involves planting native vegetation but also may require altering the landscape to restore physical features such as depressions for vernal pools that were disturbed during historic land use (Ferren and Hubbard 1996). In highly modified landscapes, where remnant native vegetation is sparse or absent, alternative approaches to selecting corridor locations—such as identifying land forms, gathering microclimate information, and examining historic land cover information if possible—may be necessary.

Identification

Once the objectives and scale have been adequately discussed and defined, steps can be taken to help identify options for a single corridor or an entire network. As discussed in Chapter 7, desirable landscape configuration should be established based on conservation principles and requirements of focal species. Selected design elements reviewed in Chapter 7 can in some cases be identified using remote imagery. These may range from 50-meter-wide (164-foot-wide) riparian corridors to mountain ridgetop linkages.

In most cases, building a GIS to integrate spatial and nonspatial information will provide an invaluable tool for corridor site identification. This is especially true for projects operating at a large spatial extent that may need to address entire watersheds, river basins, or even larger areas. Many small nonprofit organizations have used GIS on desktop computers to map and analyze relevant information. These systems are in use by communities in the most remote parts of the developing world, as demonstrated by the Masoala project in Madagascar (Kremen et al. 1999) and by conservation activists working at home who do not have any institutional support. If no computer expertise exists, one can often get a GIS project started by visiting a local school, and working with a student learning GIS (GIS is being taught in some high schools in addition to colleges). In all cases, some base maps of the landscape will be required.

In Box 8.3, we list the types of digital data that may be available and would be useful; however, this is by no means an exhaustive list and would have to be altered depending on project objectives and location. One usually starts with an underlying layer of physical geography, which is usually compiled from digital elevation models and hydrology. Climate data can be helpful for very large areas and coarse-scale planning but is rarely available at a high enough resolution to be useful for regional-scale projects

Box 8.3.

Types of Useful Digitally Mapped Data That May Be Available

Physiography:

- Digital elevation models
- Hydrography
- Temperature isoclines
- Rainfall isoclines
- Wetlands
- Soils
- Geology

Vegetation:
- Classified Landsat Thematic Mapper
- Ortho-rectified aerial photography
- Other remote sensing information
- Historic aerial photography

Development:
- Cadastral data (parcel maps)
- Agricultural land maps
- Classified Landsat Thematic Mapper
- Census data
- Land-use change models

Specialized data:
- National or regional bird surveys
- Location of threatened and endangered species
- Habitat maps
- Wildlife distributions
- Results of species occurrence models

covering local jurisdictional areas. Land cover information is often available from multiple sources at varying scales. Satellite imagery (30-meter, or 98-foot resolution) that is classified by land cover is one of the more common sources of land cover data. Often coarse-scale vegetation maps exist, and in some areas they are available for historic or presettlement periods based on historic surveys or models of the physical landscape. When focal species data are limited, land cover classes are often used as a surrogate for species and other more specialized conservation elements (Ferrier 2002).

While existing vegetation cover is often the most desirable type of information for corridor planning, we stress the equal importance of mapping developed areas. Often digital information is available for developed areas from city and regional governments. Such digital information can include existing roads, property ownership boundaries, human census data, jurisdictional boundaries, and urban service areas. These data can provide information on the level and type of development that exists across the landscape. Of particular importance are existing barriers to animal and plant movement such as roads or intensively developed areas that preclude the establishment of corridors. Cadastral data (from a map of property boundaries or parcels) is of great value and often can be linked with nonspatial data such as the number of units, land-use codes, and value of the property. This high-resolution data can be useful for estimating population densities, water use, land values, and land use.

All data suffer from accuracy problems to a lesser or greater extent. The best way to validate digitized data is by comparing mapped data with field data. A global positioning system should be used in the field, if possible, so that the field data can be accurately spatially referenced and compared with mapped data in a GIS (Behera et al. 2000). Confirming that the maps are accurate on the ground is desirable but often not practical or possible, especially for historical data.

If both present and historic information on the development footprint is available for large planning areas, then spatially explicit statistical models can be developed to forecast future land-use change to estimate future threats to unprotected habitat (see full description of these models later in chapter). Often one of the only ways to examine the influence of error associated with GIS and other location data on modeling results is through sensitivity analysis (Stoms et al. 1992), which involves modifying the original data and examining the influence of those modifications on the resulting models.

It may be important to collect additional digital data, depending on the project objectives. Much of this more specialized data will be location data, such as the mapped location of a particular threatened or endangered species. Specialized data are rarely comprehensive for the planning region, which can make it difficult to evaluate the relative importance of all sites using these data. The case of threatened and endangered species data is an informative example. These data are usually only available for sites where these species have been detected; there are no data for areas not surveyed. Also, data sources often do not record where surveys occurred but species were not detected. Not finding the species may have been due to an insufficient survey rather than the species being truly absent. The same problem exists for bird point counts and cultural resource data, two types of point location data often available in digital format. If that type of information is used to elevate the benefit of one site over another, it should be done with great caution given the incomplete nature of such data.

Another problem can arise in assuming that conserving one species, for which there are available data, will result in the conservation of other species. Research shows that in most cases, species are not good surrogates for one another (Cabeza and Moilanen 2001). This means that one species cannot be substituted for another in the analysis and produce the same results. In the field of protected area planning, surrogate measures of biodiversity are often sought to determine where to locate reserves when detailed data on the distribution of all species are not available, as is usually the case (S. A. Banks and Skilleter 2005). Using information from a suite of indicator taxa is one way to try to get around the problem. The selection of species should ideally include those representing different habitat types, structural requirements, and association with different ecosystem functions (Hilty and Merenlender 2000b; Chapter 7). This will help ensure that the selected linkages are useful to more than a single target species. However, single-species conservation is often the primary objective.

If the persistence of certain species is a vital interest, then species distribution maps and models (discussed later) may be a priority. In most cases, there are not enough field data to determine the exact distribution of a particular species. Some locations may, however, be known. Approaches to mapping the estimated distribution of species range from quick and easy to complicated and expensive. Vegetation community type maps can sometimes be used to specify the vertebrates that are associated with each habitat type. In some regions wildlife habitat relationship models have been developed in an attempt to estimate the vertebrate species associated with each habitat type (Mayer and Laudenslayer 1988). In California, for example, vegetation communities have been defined for many ecoregions (Barbour and Major 1988), and expert opinion has been polled to develop wildlife habitat models that associate vertebrate species with each community type (Garrison et al. 2000). These models remain insufficiently tested. Some research shows that they can be useful for certain species and inaccurate for others (Kilgo et al. 2002), and they may be more useful for resident than migratory species (Garrison et al. 2000). In particular, we caution against using these models for amphibians, for which our understanding is often limited, and for which existing digital mapped data are often at too coarse a scale to capture precise distributions (Howell and Barrett 1998). When little other information is available, these models can provide hypotheses of what vertebrates may occupy a site (Edwards et al. 1996). This, in addition to general range distribution maps that can be found in field guides or from museum collections, provides the quickest estimate of where particular species may occur.

Instead of relying on expert opinion, some researchers have used species presence data to model the likely distribution of species across a landscape, based on landscape features. However, these models can often require more data than are available or possible to collect quickly with a limited budget. It also can take time to develop these quantitative models. They are becoming increasingly popular (Guisan and Zimmermann 2000), nevertheless; the two most common approaches are individual-based models (Grimm 1999) and occurrence models (M. Scott 2002). Individual-based models are sometimes considered bottom-up models because they begin with individual organism characteristics and then generally model how the organism responds to a heterogeneous environment and interacts with other individuals. Traditionally, those models divide the landscape into a uniform grid. Some scientists advocate using a grid of hexagons for analysis (Noss 2003), but increasingly irregularly shaped polygons are being used to represent the actual shape of natural features.

We recommend working with actual landscape features because important linear and point features critical to corridor planning are not well represented by uniform-sized cells. A good example is a GIS-based patch model for the movements of calving elk in a short-grass prairie that used variable-sized polygons (Bian 2003). These Rocky Mountain elk rely on a riparian corridor in Cimarron National Grassland in Kansas. The elk forage in and around riparian vegetation along the corridor but require a number of seeps for water during calving. The linear riparian corridors and seeps cannot be easily represented using regular large grid cells. Elk are relatively insensitive to disturbance but do avoid high-speed roads. Various types of movement were simulated from very local movement observed during calving to movement along the corridor and at the edge of a territory. Random movement about a core location and movement in the direction of preferred land cover types was repeatedly simulated to examine the pattern of likely habitat use by calving elk. Telemetry data provided actual information on ranging patterns and were used to compare the two methods of modeling elk movement. The researchers concluded that elk movements are a combination of random movements and movements toward preferred cover. The important point is that individual-based models can be built for a small area if detailed information is known about species movement patterns and if the important features like riparian areas can be accurately mapped.

Various modeling methods are used to estimate species occurrence, the results of which are influenced by the distribution of species data used, as well as the modeling method itself (Segurado and Araujo 2004). The most popular methods currently being employed include various regression analysis techniques to determine the relationship between independent environmental variables extracted from a GIS and the presence of a particular species as the dependent variable. Elevation, land cover type, and road density are examples of independent variables that are commonly used. This approach assumes that the presence of a species is related to characteristics of the physical, biological, and built environments. Habitat suitability models using these methods for focal birds and mammals were used to identify corridors and core areas for an ecological network for Lombardy, Italy (Figure 8.1; Bani et al. 2002).

These models are sensitive to false absences that are common in species occurrence data sets (Gu and Swihart 2004). Determining that a species is actually not present at a particular location is difficult even if not detected by the survey method. That said, some models have been tested in the field and found to be robust for certain species. For example, in central southern Italy, a logistic regression model predicted brown bear (*Ursus arctos*) presence and absence correctly over 90 percent of the time, suggesting that the model would be useful for identifying critical areas for bear conservation (Posillico et al. 2004).

These occurrence models are beginning to be used to assess the risk of future land-use change on species conservation. For that purpose, a carni-

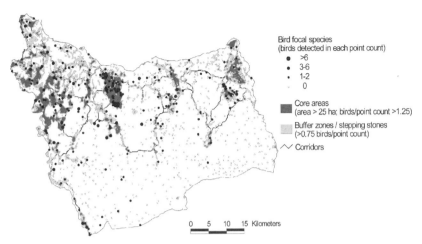

Figure 8.1. A map of the lowlands of Lombardy, Italy, showing linear corridors that were developed using habitat suitability models based on the bird and mammal location points depicted according to relative abundance (Bani et al. 2002; permission to use by Luciano Bani and Blackwell Publishing).

vore habitat suitability model was combined with a land-use change model for Sonoma County, California, to evaluate different future land-use change scenarios and possible resulting changes to species distributions (Hilty et al. 2006). A similar approach was used in Oregon to estimate the response of wildlife to landscape change. Land-use change was estimated based on historic trends, and then forecasted changes in species distribution and persistence of a few key species were presented (Schumaker et al. 2004).

It is important to remember that species occurrence models are designed to map the pattern of a species distribution, and they do not include any information on population processes. In fact, those models are poor at predicting population dynamics such as relative birth and death rates (Tyre et al. 2001). Therefore, they cannot be used for predicting population dynamics and persistence.

The most common method for modeling persistence or addressing risk of extinction is through population viability analysis (PVA), which takes advantage of life history, demography, and ecological information to quantify extinction risk (Shaffer 1981, Beissinger and McCullough 2002). These are stochastic models (i.e., they include random variables) of population dynamics, in which population viability is an estimate of the probability of a species going extinct over a specified time period. These models always include species population dynamics based on life history characteristics such as birth, death, immigration, and emigration rates. Genetic processes, including inbreeding and genetic drift, can be important in small populations (see Chapter 3) and are sometimes included in PVA. Environmental stochasticity is another important determining factor that will influence species persistence and can be modeled within a PVA. Software packages have been developed to assist in the calculations. The most widely used packages are VORTEX (Lacy and Miller 1993), RAMAS (Ferson and Akcakaya 1990), and ALEX (Possingham and Davies 1995). It is important to appreciate that the results of these models can be sensitive to the input parameters. If estimates of demographic, genetic, or environmental variables are based on a limited data set, as is usually the case, the sensitivity of the results to these variables should be tested (Mills and Lindberg 2002).

The most useful PVA models for corridor planning are spatially explicit and attempt to estimate the population dynamics at different locations in space. This approach combines traditional PVA with a GIS that includes attributes that influence the parameters used in the PVA, such as the impact of habitat type on reproductive and dispersal rates. This was done with four British mammals using a GIS that included environmental, habitat, and animal population information. These data were then combined with an individual-based population dynamics module, which simulated home range formation, individual life histories, and dispersal within the GIS-held landscape (Macdonald and Rushton 2003). An important application of dynamic PVA modeling was demonstrated by Carroll et al. (2003) to address the impacts of landscape change on wolf restoration success. To perform the dynamic PVA analysis, they connected female survival and fecundity with data on mortality risk and habitat productivity mapped for each territory in a GIS.

Population viability models have been applied extensively to the management of endangered species. However, the predictions of those models can be unreliable (Beissinger and Westphal 1998, McCarthy et al. 2003) and should be used with caution. They are most rigorously employed to estimate impacts of contrasting management scenarios rather than for predicting extinction.

One of the best approaches to corridor identification was by Vuilleumier and Prelaz-Droux (2002) and involves a relatively simple GIS model that is useful for regions where the built environment dominates the landscape, leaving habitat remnants and agricultural lands as the only option for corridor conservation and restoration. To identify possible places for an ecological network in Switzerland, the researchers incorporated digital information on roads and building areas into their model. Areas were given a relative ranking as to their likelihood of inhibiting movement of large game species, including roe deer (*Capreolus capreolus*), red deer (*Cervus elaphus*), chamois (*Rupricapra rupricapra*), and boar (*Sus scrofa*). The cumulative influence of weighted barriers was mapped to identify areas more and less difficult for animal movement. Animal movement was simulated under different scenarios to identify pathways of least resistance, such as remnant habitat and roadless agricultural land that could be restored. The approach was pragmatic and provided a flexible corridor planning tool.

Prioritization

Various types of data can be useful to identify areas for conservation and restoration that will meet the project objectives. However, the final decision depends on a wide variety of issues, some of which can be analyzed a priori. The most important components of site prioritization are the distributions of biological benefits, economic costs of the conservation action needed, and probability that the site will be lost if no action is taken. These must be repeatedly considered as resources for corridor conservation become available, because they change over time (Costello and Polasky 2004). It is important to recognize that while scientists try to provide the best prioritization methods, decision makers will want to capitalize on future political and economic adjustments that will influence which sites are ultimately conserved. Despite these limitations, it is worthwhile to understand planning components in order to make informed decisions. Like all models, reserve selection models simplify the real world. However, they are extremely useful for examining the tradeoffs associated with alternative planning scenarios that may be difficult to recognize any other way.

As this book attests, there is a great deal of focus in the conservation science literature on furthering our understanding of the biological benefits of land conservation, such as the area required for species persistence and location and habitat requirements of endangered species. In fact, the original reserve site selection algorithms were developed solely to maximize species or habitat representation across a network of reserves. Two ways of constraining site selection are minimizing the area required to protect a representative set of species or habitat types, and maximizing the representation of species or habitat types within a reserve system of a certain size. In the site-constrained framework, a species is considered protected if it is represented at any of the chosen sites. Using this approach resulted in the need to protect only twelve sites to conserve 333 vertebrate species in Southern California (Church et al. 1996).

There are software packages that use iterative heuristic algorithms to solve these problems based on the concept of complementarity, which is high when sets of sites have little or no overlap of species or habitat type (Vane-Wright et al. 1991). These algorithms search for which action will be of greatest benefit and should only be used for adaptive problem solving, as they may not result in an exact answer. An early comparison was made of reserve selection algorithms based on vertebrate data from Oregon (Csuti et al. 1997). Researchers found that linear programming algorithms were the best option for identifying the maximum number of species for a given number of sites or the minimum number of sites needed to represent all species. More recently, approaches from operations research have been used to optimally select the most desirable reserve design from complex options (Williams et al. 2004).

In addition to focusing on which corridors will provide the most biological benefit, expertise should be sought to determine the relative costs and threats of potential sites. That way, time and money can be invested in sites that require investment, so that they can continue to provide biological benefits, and thereby avoid investing in sites that by default will continue to support biodiversity. It is important to estimate only the relative probability that a site will be lost and the relative cost of conserving it compared to other sites. This means that exact figures are not as important as how the various sites are ranked from low to high. Failure to consider vulnerability or cost in prioritization efforts will result in suboptimal targeting (Newburn, et al. 2005). Economic costs of alternative strategies need to be included in order to address opportunity costs—costs of alternatives that must be forgone in order to pursue a certain action (Faith and Walker 2002).

Cost

Effective incorporation of economics remains relatively rare in the contemporary protected-area planning literature, but we highlight a few good examples. A targeting study done for Papua New Guinea demonstrated the importance of including opportunity cost into conservation priority-setting algorithms by illustrating the selected protected areas that met biodiversity targets, minimized costs, and took into account that some sites were unavailable due to conversion to agricultural land (Faith and Walker 2002). Another interesting study, which was done in South Africa, evaluated the cost of various acquisition strategies to conserve targeted lands and demonstrated that costs vary depending on the tools used to conserve private land (Pence et al. 2003). A third effort included aggregated cost data for targeting protected areas for endangered species conservation using county-level data on endangered species listings and agricultural land values for the coterminous United States (Ando et al. 1988). In that case, program costs for preserving species were shown to be significantly less when targeting also considered land costs.

The previous example used agricultural land value aggregated at the county scale to estimate land values across the entire United States. A preferred approach for corridor planning at a finer regional scale is called a *hedonic approach*. It uses observed market transactions to infer the market value of parcel characteristics (Rosen 1974). Heterogeneous land-supply characteristics may include physical land quality (e.g., slope), location attributes (e.g., proximity to urban centers), and land-use regulations and other factors influencing the returns to land (e.g., zoning). Recent property transactions are used to estimate sales price as a function of the land characteristics, in addition to any property improvements (i.e., buildings). For example, the proximity to urban center can be used to estimate the gradient in land values as one travels away from the central business district (Newburn, et al., 2006). In addition to estimating the total land value, the conservation easement value (Merenlender et al. 2004) can be estimated as the value of developable land minus the restricted-use value predicted for each developable parcel with a GIS database of parcel characteristics (Newburn, et al. 2005). For more details on the hedonic approach, Palmquist (1991) provides a rigorous explanation of the theory and application, while Garrod and Willis (1999) offer a simpler discussion with less mathematical detail.

Spatially explicit land valuation models are increasingly feasible for practical applications owing to the availability of parcel databases and advancements in GIS technology. Parcel records, collected for tax assessment purposes by local and state governments, provide detailed information on property sales, existing-use value assessments, land use, and other characteristics. These records can be linked to a digital parcel map and integrated with site-specific characteristics to estimate the developable land value per acre as a function of parcel characteristics.

Vulnerability

Vulnerability or the likelihood of future land development or conversion also needs to be considered. Some areas of the landscape will remain protected because they are not suitable for development or may be at risk in a long time horizon compared to other sites. If only conservation benefits and cost were considered, then it would be assumed that all sites are equally likely to be lost in the future, which is rarely the case. The expected probability of land-use conversion often increases as a function of the value of developable land. This is usually true when the external threat to the site is development, agriculture, or logging. The positive correlation between the probability that a site will be converted and its cost means that low-cost parcels typically have a low likelihood of future conversion. If the relationship between vulnerability and cost is not accounted for, low-cost sites will be selected even if they are not threatened. For that reason, it is important to consider whether a site needs investment to protect or restore it, in addition to how much it would cost to do so (Newburn, et al. 2005). Therefore, it is important to estimate the probability that a site will be vulnerable to land-use change (K. Wilson et al. 2005). Typically, ad hoc ranking or rule-based classification is used to formulate a proxy for vulnerability (Abbitt et al. 2000, Pressey and Taffs 2001).

A better approach for finer-scale decision making is to actually estimate land-use change as a function of the underlying biophysical and socioeconomic characteristics, based on sites that have previously been converted (Chomitz and Gray 1996, K. Wilson et al. 2005). Consider a simple land-use change model constructed with respect to a set of developable parcels observed at two time periods. For each developable parcel, there is a binary outcome: remain in the initial developable land use (e.g., forest habitat) or be converted to a more intensive type of land use (e.g., housing). Mapped biophysical and socioeconomic characteristics derived from a GIS serve as explanatory variables in a logistic regression to estimate the relative probability of each land-use alternative. For example, forest conversion to agricultural use will be more likely on areas with suitable soil quality, slope, access to water or precipitation, and access to markets. Coefficients from the logistic regression then may be used to predict the relative probability of land-use conversion for remaining developable sites.

This type of land-use change model was developed to examine the threat of residential development to extensive agricultural land and open space in Summit County, Colorado (Theobald and Hobbs 1998). This study compared the results of a model using regression analysis with those of a transition probability model that assumes the neighboring areas influence the transition probability of the central area in the same way that cellular automata models do (Zhou and Liebhold 1995). As we might expect, the latter approach produces more clustered patterns of development, which in this case better represented the observed patterns of development. This example demonstrates two approaches to modeling land-use change using available data, but a broader discussion of land-use change models is available (e.g., Agarwal et al. 2002, Berling-Wolf and Wu 2004). The important point is that land-use change models are a better way to examine vulnerability compared to more ad hoc estimates.

Future Directions

Recently, the limitation of models for reserve selection over a several-year planning horizon has been demonstrated. In fact, because optimal reserve selection models generally do not consider changes that can occur over time, they often are outperformed by a simple opportunistic approach using principles of conservation biology, such as prioritizing sites according to species richness (Meir et al. 2004). These limitations have been recognized by researchers who are attempting to make the models dynamic over time, so as to include future changes that are likely to occur to species, sites, and the value of land (Newburn, et al., 2006). Another important limitation that has only recently been addressed in modeling reserve selection is the availability of sites for acquisition or long-term management agreements (Meir et al. 2004). The availability of sites for protection or restoration does impact the implementation of any corridor plan. Site availability often depends on the landowner, especially in a privately owned landscape, and it can be difficult to predict. So we see that in many cases, rather than presenting a solution, reserve selection models primarily inform us about the process and how sensitive solutions are to the model and variables used.

Several tools and services to assist conservation organizations have been developed. The Orton Family Foundation's CommunityViz program (www.communityviz.com) has created GIS-based software packages that allow communities to visualize and quantify the impacts of alternative conservation and development scenarios. NatureServe, a spin-off program of The Nature Conservancy, will soon release a decision-support system designed to map and integrate scientific data for conservation prioritization (Stoms et al. 2005). While both tools are supported by extensive digital databases, they provide solely ad hoc methods for assessing economic and land-use change components. Spatial models for these components must be constructed separately or imported from independent sources and may not be readily available for many regions.

Assessing the Potential Utility of Priority Corridors

Most people recognize the importance of field visits prior to any conservation action in an area. The type of data that may be useful to collect at a particular site to estimate its utility as a potential corridor will depend on the initial goals for the site, resources, time available, and site access. The objectives of site assessment fall into four categories: (1) habitat type and structure, (2) condition of the site, (3) presence of target taxa, and (4) limiting factors.

Here we review some methods that can be useful for quantifying site characteristics of various corridor options. The process usually starts by classifying the existing or historic habitat type. A suite of species that are associated with the habitat type can be hypothesized to be present. Indicator taxa can be identified and surveyed to assess site condition. If the conservation of a particular species or community is the primary reason for protecting or restoring habitat linkages, whether that species or community is using the area can sometimes be determined through field surveys. The nature of some species may make their presence difficult to rapidly detect. In such cases, assessing the habitat type and condition may be a necessary surrogate for determining species occupancy. Equally important to measuring habitat type and quality is determining any potential limiting factors that may prevent or reduce the utility of a particular site.

Preliminary baseline information about potential corridor locations should include physical, biological, and social information. Some physical information, such as the location of the site within a particular drainage area, can often be determined from existing maps that may already be part of the project GIS. However, some mapped information may be too coarse to accurately portray a particular site. Therefore, site characteristics such as slope, aspect, geology, and soil types may need to be confirmed with field measurements. These variables influence the type of vegetation that may currently or historically be found at the site. If the site is large, it may be necessary to select a set of sampling points where data will be collected. It is important that the sample sites represent the breadth of variation found across the area under consideration. The best way to ensure this is to select sampling sites based on the mapped information, even if it is coarse information. For example, if multiple soil types and changes in stream gradient occur, randomly selecting sample locations evenly across the range of soil types and classes of stream gradient would be best. This is referred to as *stratified random sampling* and is often used in ecological studies (Lookingbill and Urban 2004).

While digital maps of land cover can sometimes provide coarse-scale information on vegetation types, that information is never completely accurate. Therefore, it is useful to determine the habitat type of a particular site in the field. This is often done using existing classifications, which define habitats according to dominant plant species and associated species. There is much debate in the literature about the utility of habitat classifications that rely on defined climax community types (J. E. Cook 1996) because ecosystems are dynamic over time and may not reach equilibrium at a single climax community type due to perturbations and site and global conditions. However, it is helpful to obtain existing plant community classifications and determine which community types are represented within the corridor sites of interest.

In addition to habitat type and likely occupants, the overall condition of a site may be especially important to determine if the focal species are intolerant of disturbance. Evaluating the ecosystem health or condition of a particular site is often accomplished by identifying the presence of indicator taxa. Indicator taxa are species or higher taxonomic groups whose parameters, such as density, presence or absence, and infant survivorship, are used as proxy measures of ecosystem conditions. They have been used to evaluate many aspects of environmental condition, including toxicity levels, biodiversity, target taxa status, and ecosystem health.

That no single taxa can accurately reflect ecosystem health is well understood (Hilty and Merenlender 2000b); nevertheless, we often need to rely on indicator taxa to measure the ecological integrity of certain areas (Carignan and Villard 2002). For some systems, indices of biological integrity have been developed that rank site condition based on the composition of the biological community and often take advantage of existing information on local species sensitivity to disturbance. For example, the index of biotic integrity (Karr 1981), developed to assess degradation in streams, uses an array of ecological measures, one of which is indicator taxa (Fausch et al. 1990).

There is a good deal of information in the literature about how best to select indicator taxa (e.g., Landres et al. 1988, Noss 1990, Soulé 1985). After reviewing those criteria, we recommend that the most logical approach to

Box 8.4.

Decision-Making Framework for Selecting Indicator Taxa

Step 1. Decide what ecosystem attributes indicator taxa should reflect.
Step 2. Make a list of all species in the area that best satisfy the following baseline criteria:
- taxonomy is clear;
- ecology and biology are well studied;
- tolerance levels for human perturbations are known;
- correlation to ecosystem changes is established.

Step 3. From the preceding list, retain species that best meet the following criteria:
- provides early warning and functions across a range of impact levels;
- has detectable trends;
- has low variability;
- is a specialist.

Step 4. Remove species that may respond to irrelevant changes occurring outside the system of interest.
Step 5. Use only those species that can be easily detected and monitored with available funds.
Optional step. Reduce the list further by selecting taxa in the list with widespread distributions or that represent other agendas of interest.
Step 6. Select a set of complementary indicator taxa from different taxonomic groups so that all selection criteria are met by more than one taxon.
Source: Hilty and Merenlender 2000b.

selecting indicator taxa can be summarized in the steps listed in Box 8.4. The search should start with species that have adequate baseline information including life history traits, taxonomy, and tolerance to disturbance (Hilty and Merenlender 2000b). Specific niche characteristics should also be considered for each potential indicator taxon. Another variable commonly noted in the literature is that selected indicator taxa should not greatly fluctuate in population size under natural conditions, which could make detecting extrinsic impacts to population difficult to detect. In addition, the life history of the selected indicator taxa should be such that they will be able to both provide early warning and be effective over a wide range of stresses (Kelly and Harwell 1990, Soulé 1985). Finally, their habitat or trophic specialization can be important. Species with high community dominance (influence on community structure and function) will often meet many of these criteria. For example, monitoring common species of zooplankton in aquatic communities may quickly reveal major changes in the health of the system.

Other considerations proposed for selecting indicator taxa include cost effectiveness and ability to detect and quantify changes in numbers and condition (Pearson and Cassola 1992). Inadequate sample size, lack of statistical power, or inability to detect the difference between changes caused by normal environmental variation and those induced by human impacts can be problematic when trying to determine the level of disturbance at a particular site. Finally, use of a set of complementary indicator taxa in which each selected taxon can satisfy multiple criteria for inclusion is critical.

Negative indicators, such as aquatic fly larvae (Diptera), brown-headed cowbirds (*Molothrus ater*), and Norway rats (*Rattus norvegicus*), are potentially easier to find, quantitatively measure, and manipulate. Negative indicators are often generalists and can provide data over a larger range of stresses, although, as with all indicator taxa, changes in the taxa need to be correlated to changes in the community (Landres et al. 1988). These species can be very useful in estimating the general level of disturbance that exists at a site. Also, species that are adapted to disturbance can negatively interact with native species and potentially limit the utility of a wildlife habitat corridor (Chapter 6).

Direct on-site monitoring can help estimate the presence and relative abundance of particular species. Local people who rely on extracting natural resources for their livelihood are often skilled at detecting the presence of particular species (Merenlender Kremen et al. 1998, Poulsen and Clark 2004). Some wildlife biologists have spent years in the field tracking and learning about wildlife and may be helpful in identifying signs that animals leave behind in the form of scat, tracks, and other artifacts. In some cases, local naturalists may be able to identify birds by their song and provide useful site-specific bird lists. From these sources, information on focal taxa or indicator species can sometimes be extracted. The same is true for flora, as local experts are often members of native plant societies and can provide valuable plant lists, including information on endangered species and vegetation assemblages. Track plates and other methods for detecting wildlife tracks, such as smoothing the ground surface or observing fresh snow and mud, can sometimes be effective but often require an expert to interpret the tracks (Loukmas et al. 2003, K. A. Mooney 2002).

Dogs have recently been used effectively to detect wildlife scat in a variety of habitats. For example, the use of scat detection dogs in monitoring San Joaquin kit fox (*Vulpes macrotis mutica*) improved the efficiency of off-road field surveys by 900 percent (Meadows 2002). For this reason, the use of dogs to study wildlife populations has become popular among researchers (D. A. Smith et al. 2001).

Using remotely triggered cameras offers several important advantages, particularly in narrowly vegetated corridors (Hilty and Merenlender 2004) and underpasses (Ng et al. 2004), where they do not require baiting and therefore represent a passive monitoring method. Cameras in some cases detect more species than track plates at the same location (Hilty and Merenlender 2000a; Figure 2.5). Photographed animals can easily be identified to species, and sometimes individuals can be distinguished from one another even by nonexperts (Karanth and Nichols 1998).

This information provides some guidance on how to assess the utility of potential corridor sites for wildlife conservation. Implementing these methods requires some time and financial investment but may be necessary to justify corridor conservation, particularly for species recovery.

Limiting Factors

The potential pitfalls that corridors can present to some species have been discussed in Chapter 6. They should be avoided when possible, particularly if they might apply to focal species in need of conservation. A site may be generally suitable for the desired purposes but have limiting factors that will prevent the success of the project. These factors may be physical, biological, or social. It is important to try to identify these at the earliest stage possible to avoid investing in an effort that will not produce the desired results. A physical barrier such as a dam, road, or natural cliff may prevent animal movement through a particular corridor. A particular invasive species or urban adaptor may prevent native species from using or reproducing in a corridor (see Chapter 6). Some landowners may be unwilling to accommodate a corridor project on or adjacent to their land. The sooner such limiting factors can be identified, the better.

While digital data on the built environment are extremely valuable, they, like the biological data, may not be at a fine enough resolution to determine some site-specific limiting factors. For example, while paved roads are often available on digital maps, dirt roads, which can also present problems for corridor integrity, are less likely to be mapped. The location of surface water throughout the year is important to consider on any site. While streams and wetlands can attract some species, they can inhibit others. Small water sources such as springs and seeps are important natural resources but are rarely well represented on digital maps. Also, considerations of the likelihood of being able to restore the selected site need to be addressed. For example, water sources may have to be identified to irrigate plantings, and regular access must be possible.

The importance of species interactions is often underappreciated in determining suitable habitat or potential corridors for individual species. In the fragmented landscape of Southern California, for example, coyotes (*Canis latrans*) were found to be competitively dominant over gray foxes (*Urocyon cinereoargenteus*) (Fedriani et al. 2000). Given the potential for native species to competitively exclude one another, and all the potential negative interactions of exotic species, such as direct competition, predation, and disease transmission, it is important that information on other species that may limit the success of focal species be collected. More examples of negative species interactions that can occur in corridors can be found in Chapter 6. The more information that can be collected in the planning process regarding potential limiting factors, the less likely a corridor project will fail to meet its stated goals.

In summary, determining the type, quality, and potential utility of possible corridors, as well as prioritizing sites for conservation, should be done before long-term investments in conservation and restoration are made. Measuring these factors will also provide baseline information for continued monitoring and adaptive management of the corridor or network.

CHAPTER 9

Protecting and Restoring Corridors

The previous chapter discussed the importance of systematic conservation planning and described the tools used to identify priority sites for conservation. Each site will have its own peculiar ecological, physical, economic, and social circumstances, which will dictate the steps required to secure the site as a functional corridor. Protecting and restoring corridors require some of the actions that we discussed in Chapter 8, including identifying specific goals and collaborating with stakeholders. Also, ecological monitoring in and around the proposed corridor site should continue to ensure the desired outcomes.

In this chapter, we briefly describe general strategies that should be part of any implementation project, then examine the tools used to conserve land for connectivity. In particular, we review the recent proliferation of incentive-based conservation tools, including purchasing a partial or full interest in land. Once land is acquired, restoration is often necessary. Ecological restoration is too vast a subject to include in depth in this book, so we cover general points to consider. In the second half of this chapter, we provide project case studies implemented at varying spatial scales to illustrate principles discussed in this chapter and throughout the book. We attempt to present less well-known cases rather than more familiar examples from the conservation science literature, which, in many cases, are referenced in other chapters. The cases we present are based primarily on informal sources such as Web pages and through personal experiences and contacts.

General Strategies

As in the case of the planning process, public participation is usually required to implement a corridor project. In general, place- or community-based groups (Chapter 8) build partnerships and work together toward increasing connectivity through corridor establishment. Many individuals and communities around the world volunteer to fence, plant, and care for wildlife corridors across private and public land. Landowners often obtain financial assistance from publicly funded programs to help with a particular project, and they usually more than match those funds with their own time, equipment, and land resources. From the literature and from identifying commonalities among corridor projects, we recommend the following general strategic components for corridor implementation (Rapp 1997):

- *Community relationships.* Develop close working relationships among local citizen groups that are interested in land and water resources and may be organized around professions such as farming or fishing or interests such as bird watching or watershed health. Developing such relationships is an ongoing process that is essential for any project. In some cases, a corridor project will be initiated at a higher institutional scale, such as by a large nonprofit conservation organization or a public agency. In that case, it is most important to involve local community members where the project will be implemented, and local interest groups are a good place to start.
- *Human resources.* Establish contact with resource agencies and other sources of expertise and possible funding early in the process. Many of these agencies and organizations produce leaflets and other educational materials valuable for community education, and they can be invited to local fairs or field days to present information. These experts can also help identify private and public funders to help implement land conservation and restoration.
- *Diversity.* It is important to understand the breadth of ideas and practices that may be important to adopt in a corridor project. Local and professional contacts will add to the diversity of objectives and strategies for restoration. Often, indigenous people have effective methods of working the land for the benefit of nature and community. Decisions about what to conserve and restore and how to do it should be informed by the diverse opinions represented in a wide

range of partnerships. This approach will enhance the potential benefits of any project.

- *Education.* Ensure that community education programs that illustrate why conservation and restoration are important for local flora, fauna, and human communities are in place. Public schools, local radio, print media, and businesses that employ large numbers of local people are all good places to distribute materials that provide justification for a corridor project. These materials should give local people a sense of place, which can lead to a feeling of pride in restoring and maintaining a healthy environment. Many public education programs also post signs to inform the public about the resources in need of protection and to minimize negative impacts from local people in and around the corridor. Depending on the project goals, this may include "No pets" signs, information about invasive species, and road signs to caution drivers about the presence of certain wildlife species (Figure 9.1). Some European countries are testing collision avoidance systems with game detection capabilities that warn drivers of the presence of large animals to reduce collisions. Box 9.1 discusses an example of where signs and road closure were used to protect an amphibian population.

- *Ecosystem approaches.* Try to maintain and restore the self-sustaining ecosystem processes in addition to native species, so that important functions can be maintained or recovered. This will require consideration of the larger landscape, including hydrological and geological processes, as well as local sources of pollutants, even when working on a small-scale project at a particular location. A long-term perspective is also needed because, for example, although trees can regenerate over several years, rivers may take much longer to change channel and floodplain formation.

- *Expectation of disturbance.* It is important to remember that natural and human-induced disturbances can affect the establishment, maintenance, and effectiveness of a corridor. A rare large disturbance event could occur shortly after an active restoration project is implemented, resulting in plant loss or physical damage, for example. Some actions can be taken to prevent losses, such as posting signs, building fencing, and securing treatments for extreme weather events. For long-term success, a system of protected areas and linkages between them needs to be resilient to disturbance that may be unpredictable.

Box 9.1.

Caution: Newts Crossing

In 1988 there was a meeting of the East Bay Regional Park District Board of Directors to receive public input into their developing land-use plan for Tilden Park (Alameda and Contra Costa counties, California). One citizen asked if anything could be done about all of the newts (*Taricha*) that were killed each winter as they crossed South Park Drive to reach their breeding stream, which followed much of the road. South Park Drive is a major road through the park and was increasingly being used by commuters. The board's response was to the effect that newts had been killed on that road for years and were doing just fine. The exchange initiated much discussion, however, and the issues raised were transmitted for comment to Dr. Robert C. Stebbins, herpetologist at the Museum of Vertebrate Zoology, University of California, Berkeley. He agreed that there was a problem and suggested that a study be performed to evaluate the level of newt carnage. This idea was accepted, and the board approved a plan on July 19, 1988, to close the road (2 kilometers, or 1¼ miles) following heavy rains, effective in the fall of 1988 (Figure 9.1).

Road closure was met by a confused and sometimes angry public. Moreover, the policy of opening and closing the road all winter as rainstorms came and went added to the frustrations of motorists and the workload of the park staff. They did their best to educate drivers about the project, and in the autumn of 1989 Dr. Stebbins and a newly employed

Figure 9.1. Sign indicating that road is closed for newt crossing, Tilden Regional Park, Alameda County, California. (Photo by William Z. Lidicker Jr.)

naturalist (Jessica Shepherd) began their research project. They marked off South Park Drive at 10-meter (33-foot) intervals, and regularly surveyed for newts, both alive and squashed. They found that newts were crossing the road all winter, rain or shine. In 1992, the park's supervising naturalist presented their results to the board, complete with a large jar of flattened newts. The board agreed to have the road closed for the entire rainy season, November 1 to April 1. Public reaction was mixed, but at least now drivers knew when the road was to be closed and could plan accordingly.

This unusual management tactic to help newts move back and forth between their breeding stream and upland nonbreeding habitat continues to the present time. An unanticipated benefit of the plan is that when the road is closed, it becomes a favorite route for hikers, dog walkers, cyclists, and even equestrians. The public has accepted it as the normal course of the annual cycle in Tilden Park.

- *Adaptive management.* Corridor establishment and conservation are adaptive processes that require a continual feedback loop in order to adjust to changes in the sociopolitical arena, new funding opportunities, unforeseen disturbances, and new collaborative partners. Monitoring is essential for adaptive management and is the only way to effectively evaluate project outcomes and the ecological and social benefits of a project. Therefore, we strongly recommend collecting baseline ecological and social information about a project area and then systematically quantifying changes at an appropriate temporal scale, as discussed in Chapter 8. Unfortunately, the monitoring of ecological and social change is often unaddressed and underfunded.

Land Stewardship and Education

Voluntary efforts by landowners to increase sustainable land-use practices and conserve natural and agricultural resources represent an important mechanism for conserving corridors. These efforts represent the most informal types of conservation agreements, but they can affect far more land than incentive-based agreements will ever be able to address. Land stewardship programs encourage landowners to implement best management practices and conserve resources through education and community involvement. These programs are often supported by government agencies that provide landowners with educational materials. Working with farm extension agents and sharing knowledge among local landowners are also effective ways of extending information on best management practices. Landowners participate in voluntary programs for a multitude of reasons that include increased access to information and local expertise, and public recognition for those practicing sustainable methods.

Many countries have land care programs that encourage landowners to join community-based sustainable land stewardship groups. In Australia, for example, the Department of Agriculture, Fisheries and Forestry supports Landcare, a voluntary community group movement focused on improving natural resource management. Approximately four thousand groups operate mostly in rural Australia. The Australian government plans to provide AUD$159.5 million for Landcare from 2004 through 2008. Complementing this government support is a private nonprofit group called Landcare Australia, which raises funds through corporate sponsorship to support land care community groups. A different effort in Western Australia, the Fitzgerald Biosphere Project, has "produced considerable benefits for corridors in the region" (Bradly 1991). This was done through innovative farm plans that included planting native species in agricultural belts to serve as wildlife corridors. This example demonstrates that sustainable land use can be promoted by community organizations such as land care groups and indigenous cultural groups. These community organizations are often the best place to start working on ecological networks across private landscapes.

The demand for change, ideas on ways to change, and mechanisms to implement change all need to arise at least in part from the local people themselves. Building this momentum from the ground up takes time and therefore requires a long-term commitment. Getting people to change their behavior and adopt new practices can be a slow and daunting task. Relying solely on education programs may be insufficient in areas where there is a high turnover of landownership.

Children's education programs may provide long-term returns for conservation. Macedon Primary School in Victoria, Australia, started the Slaty Creek Underpass Wildlife Project. The project was sponsored by the state roads management agency, and the children monitored wildlife use of a road underpass for an entire year. They worked with scientists and artists to help document the use of the underpass by wildlife (Abson 2004). The results were a better understanding of the species that use the underpass and an exhibition to communicate the findings of the project to the community. This is a wonderful example of how conservation action can engage children and teach them about how to better manage human and natural systems.

Indigenous people are the stewards of much of the world's remaining open space, and the success of many corridor projects will rely on them. With indigenous cultures rapidly disappearing (Dalby 2003), there is increasing concern that some conservation measures have contributed to further marginalization of indigenous groups. Care should be taken to ensure that conserving biodiversity does not inadvertently result in another form of oppression for indigenous people.

Conflicts between parks and local people have a long history and can be attributed in part to the top-down approach generally used to establish protected areas (Peluso 1993). That approach had negative consequences for the people and for the parks. Community-based conservation started in Africa as a backlash against regulating wildlife and land protection from the top down and is supposed to devolve natural resource management to local communities (Hulme and Murphree 1999). That objective requires more than considering local communities as stakeholders in the decision-making process. It means listening to the experiences of the local elders in particular and trying to understand how they and their ancestors interacted with their environment and thereby managed resources for sustainable returns.

There is still a great deal that needs to be done for conservation interest groups and governments to truly integrate indigenous knowledge into land management and to empower local people with control over land and wildlife (Sefa Dei 2000). One of the greatest challenges is recognizing the diversity that exists within a single people or village and figuring out how to include indigenous beliefs and histories in project development without simplifying them and weakening their integrity. We have a long way to go before we can define an effective approach to working with local people to successfully conserve biodiversity, but we do know that empowering local people in the process is a necessary first step.

Purchasing Land for Conservation

Stewardship, education, and government policies all influence how land is conserved. Increasingly, however, private organizations and governments are purchasing interest in land to protect natural habitat and compensate landowners for opportunity costs associated with necessary changes in land use. The most common mechanisms for doing this are (1) public or private acquisition of land and establishment of a park or protected area; (2) partial purchase of land rights that places limits on the use of the land; (3) land management agreements that facilitate the protection of resources; and (4) purchasing concessions. These actions may occur sequentially on the same piece of land. For example, a landowner may enter a long-term management agreement to fence livestock out of creeks on a ranch and then sell or donate a conservation easement to prevent extensive development on the same property. These tools are most effective when combined with other tools such as education and regulations. For example, landowners may be motivated to sell interest in their land for conservation to reduce their property value and taxes. Changes to tax policy can greatly influence these incentives and potentially reduce the utility of this approach to private land conservation. Therefore, most mechanisms used to conserve land rely on a combination of regulations, incentives, and education. These interdependencies mean that the options for incentive-based conservation can change quickly over time.

Multiple public and private partnerships and a combination of funding sources are used to finance incentive-based conservation actions. In some cases, a public agency may not be able to act quickly enough to obtain a desirable piece of land, so a nonprofit organization may purchase land from a private individual and then resell it to the public agency once the agency is able to complete the transaction. Despite the perception that incentive-based conservation is privately supported, much of the funding is public, including, for example, federal, state, and local tax breaks. Additional public funds come from federal land management organizations, money paid for mitigation of environmental impacts, direct local taxes for open-space preservation, government general funds, and special bonds. Any transaction may combine a blend of funding from any of these sources, resulting in complex partnerships that make determining accountability difficult.

Purchasing Public Parks and Private Reserves

Reserves owned entirely by the private sector, the government, or a mixture of the two may provide the most security for the natural resources at risk. However, managing parkland can be very expensive, and so some public agencies hesitate to acquire land even when the resources for the initial purchase are available. Often the resources required for land management to protect a reserve are inadequate, leaving the resources at risk. Parks that are not adequately protected are often referred to as "paper parks," as they appear on paper but do not necessarily protect against disturbance such as forest clearing and harvesting. The effectiveness of a park is related to its level of enforcement and other protective measures; when funded, parks can increase the protection of biodiversity (Bruner et al. 2001). Also, most public land is open to recreation and other multiuse purposes. Those uses may limit the capacity of reserves to sustain wildlife populations. Several studies have shown that wildlife avoid or flee from hikers, bikers, and equestrians (Miller et al. 2001, A. R. Taylor and Knight 2003) and that various types and intensities of recreational use have differing impacts on sensitive species (Papouchis et al. 2001, A. R. Taylor and Knight 2003). The establishment of recreational trails can restrict the spatial distributions of wildlife populations (Fairbanks and Tullous 2002), leading to the loss of otherwise suitable habitat and access to resources. If parks are part of a corridor plan for focal species conservation, species' sensitivity to humans and pets should be considered prior to assuming that publicly owned and managed land is the best solution.

Conservation Agreements

A conservation agreement is a voluntary contract between a landowner and the holder of the agreement, which is often a public agency or land trust. This is generally a flexible, voluntary agreement that is negotiated with the landowner, and therefore a wide variety of agreements exist. Such contracts specify a conservation interest in the land and can impose land management requirements or use restrictions for the protection and enhancement of natural resources. The resources usually include unique natural habitats, prime farmland, or species of concern. A conservation agreement can be for a certain length of time or can be permanent and registered on the land title. One of the most prevalent types of conservation agreements used in the United States is the conservation easement, which includes the permanent transfer of development rights. In Australia, permanent conservation agreements are sometimes negotiated between indigenous communities and the government concerning management of aboriginal land.

In most countries, the law dictates which type of institutions can enter a conservation agreement with landowners. These are generally public agencies and nonprofit conservation organizations, which include conservation land trusts. The Land Trust Alliance in the United States defines conservation land trusts as any organization that acts directly to conserve land that is independent of the government. More than 1,500 land trusts have protected more than 3.7 million hectares (9.3 million acres) of land, primarily through donations and acquisitions of full titles or conservation easements, which restrict harmful land uses (Land Trust Alliance 2005).

Conservation land trusts are often presumed to operate independently of government, but in fact that is rarely true. They rely on tax relief from federal, state, and local governments and often seek public grants to support their activities. There is also a small but growing number of government agencies that function similarly to land trusts. A successful example of that is the Sonoma County Agricultural Preservation and Open Space District in California, which is funded by a local sales tax. New Jersey, Vermont, and Maryland are among the U.S. states that run their own conservation land trust operations. Conservation land trusts can range from very small nonprofit organizations without permanent staff to large international organizations such as The Nature Conservancy. Characteristics of the land trust such as size and mission will influence the projects that they engage in, the level of monitoring conducted, their long-term persistence, and the ultimate conservation outcomes.

Similar institutions have formed in other countries for the same purpose. The World Land Trust based in the United Kingdom has helped purchase and protect over 121,405 hectares (300,000 acres) of threatened wildlife habitats worldwide since its inception in 1989. Many conservation land trusts outside of the United States are public agencies that disperse funds for land conservation agreements. An example of this is the Australian Heritage Trust, which was established by the Australian government in 1997 to help restore and conserve Australia's environment and natural resources. Some conservation trusts that are started by governments also receive private support for acquisitions. An example of this is the French Conservatoire du littoral (http://www.conservatoire-du-littoral.fr), a public coastal protection organization that receives private donations for land and marine conservation. This organization acquires land for resource protection primarily through private agreements. The bottom line is that there are a large number of these institutions, especially in the United States, and they are often integral in implementing corridor protection and restoration projects. For a thorough examination of the history and practice of conservation trusts, read *Conservation Trusts* by Fairfax and Guenzler (2001) and *Buying Nature* (Fairfax et al, 2005).

Conservation Easements

The conservation easement or covenant is one of the primary types of agreements that conservation land trusts use to protect environmental resources. Under conservation easements, land is retained in private ownership, and a land trust or government agency acquires nonpossessory interest in the property, restricting use for the preservation of natural resources, agriculture, or social and cultural amenities. The restrictions stay with the title of the property and are therefore transferred with the property if the land is sold. In return for donating or selling a conservation easement, a private landowner receives a reduction in taxes because the overall value of the property is presumed to be less once some nonpossessory rights are removed. However, in some cases, the land value can increase due to the benefits of conserving environmental resources such as native trees. Elevated housing prices in neighborhoods adjacent to protected properties have demonstrated the public's willingness to pay more for open-space amenities (Standiford and Scott 2001).

Between the early 1980s and 2000, over 2.5 million hectares (6 million acres) had been placed under conservation easements in the United States (Gustanski and Squires 2000). These easements are often considered a win-win solution, because private citizens retain ownership, while the public pays to a greater or lesser extent for conserving natural and agricultural resources associated with the land. However, it is unclear what exactly is being conserved with conservation easements and who is benefiting. There is much to be done in examining the conservation outcomes of conservation land trusts (Merenlender et al. 2004).

Mounting public criticism of conservation easements as the solution to private land conflicts may eventually result in the pendulum swinging back toward regulatory approaches. This is because some landowners are taking advantage of the tax breaks that conservation easements provide for land that is not developable or does not provide sufficient public trust benefits. People raise their eyebrows when they hear about golf course owners receiving large tax breaks for donating easements for questionable "wetlands" within the golf course development. Criticism also stems from the fact that conservation easements do not generally provide public access to the land for recreation and other uses. Unlike land-use regulations, decisions regarding private land incentives are usually made in agreements between private organizations or the government and individuals. Since these agreements are attached to property titles, the public is often not explicitly notified of the transactions, even if public funding is used to secure the protections, as is frequently the case. Nor are communities generally aware of potential tax revenue losses until it is too late. The effectiveness of these easements is also questionable because violations may not be detected, landowners are usually not prosecuted, and easements can be invalidated in a court of law.

Management Agreements

Long-term management agreements are another type of arrangement that is commonly used by resource agencies to protect natural resources on private land. Management agreements are voluntary agreements between a landholder and another party, detailing the use and management of the land for a set period of time. These agreements outline land management responsibilities and are not binding if the land is sold. They are extremely flexible documents that can include land management activities such as grazing regimes, fencing, riparian protection, tree planting, and restrictions on resource extraction. The most widely known program that issues these types of agreements is the U.S. Department of Agriculture's Conservation Reserve Program. It provides cost-share assistance to establish conservation practices on agricultural land by entering into ten- to fifteen-year contracts with landowners. Europe's agri-environment programs (Chapter 1), designed explicitly to conserve natural resources, are another good example.

Often, management agreements are simply between a landowner and those who manage the land, for any number of purposes. But, increasingly, incentives are provided to landowners to enter long-term management agreements designed to protect and enhance natural resources. For example, a nonprofit conservation organization based in the United States called Ducks Unlimited offers financial incentives to landowners willing to manage their land for waterfowl and other wetland wildlife under a ten-year agreement. Such payment programs are also becoming an increasingly popular tool for biodiversity conservation in tropical countries. The Costa Rican National Forestry Financial Fund pays farmers, through non-governmental organizations, to provide carbon sequestration, which also protects hydrological services, biodiversity, and scenic beauty (Castro et al. 1998). The role of financial incentives and surrounding issues such as property rights, institutional arrangements, and monitoring are discussed in Ferraro (2001). Incentive-based conservation may be on the rise because more indirect methods, such as long-term investments in development, frequently fail, and direct payments can result in immediate conservation (Ferraro and Kiss 2002), although the long-term sustainability of this approach remains untested.

Because conflicts may arise among contracting parties, it is wise to address as many issues as possible in advance and include in the agreement a dispute resolution mechanism that does not require litigation. For example, the two parties should consider writing into the agreement that they will go through mediation with an identified mediator to resolve conflicts if they arise. As with most legal contracts, if either party breaches the contract, the other has the right to sue to recover damages; however, court costs may prevent such enforcement. More likely, noncompliance will lead to agreement termination. Agreements can also be terminated if both parties agree on a premature termination or, of course, when the fixed period of time expires.

Management agreements are often more detailed with respect to what must get done than permanent agreements that transfer with the land title and cannot be renegotiated. In other words, landowners are often willing to follow more specific guidelines for managing land and enhancing resource protection under agreements that are not binding in perpetuity. The increased specificity of these agreements often includes precise areas that need to be protected or managed for the protection of species or natural habitat. That is why these are particularly popular for wetland management. In summary, these agreements can be more precisely tailored to the conservation organizations' available budget, the landowners' willingness to accommodate changes in land management, and the resources of concern.

These long-term agreements have advantages for both landowners and conservation organizations. Landowners do not have to forgo the opportunity costs associated with the future sale of their land. The costs of the agreement are fixed, and therefore the conservation organization does not incur any unforeseen costs or other problems often associated with land ownership and management. There are also potential disadvantages of these agreements for both parties. The investment in negotiating these agreements and the initial activities to improve the condition of the land can be lost if the land is sold. Very specific management guidelines may not be realistic in a rapidly changing environment. For example, restrictions on grazing regimes may be dependent on sufficient rainfall in any one year. Extreme weather in combination with management constraints could make livestock production at the site impossible in the future.

Conservation Concessions

A conservation concession involves purchasing a lease that generally is sold by a government to a company for extracting resources through forestry or mining. Conservation organizations have started to provide periodic payments as part of a fixed-term lease agreement in order to prevent the removal of natural resources in developing countries. This mechanism has the potential to provide important protection from deforestation for the term of the lease (generally, fifteen to forty years) and possibly longer if renewed. The most widely publicized agreement so far was between Conservation International and the Guyana Forestry Commission for an 80,000-hectare (197,600-acre) forestry concession called the Upper Essequibo Conservation Concession in southern Guyana. In addition to protecting biodiversity, conservation organizations claim that conservation concessions provide revenue for development at the local level. However, how much money will trickle down from the government to local people from conservation concessions remains to be seen. This is an important point, since forestry companies often provide jobs, roads, schools, and other infrastructure for local communities.

The same issues for prioritizing land conservation that we discussed in Chapter 8 should apply to these types of agreements, as well. Namely, in order to determine the value of these agreements, the conservation benefits and relative threat level should be assessed, in addition to cost. Also, the unintended consequences of protecting land for local communities and other natural areas need to be explored. For example, in the Guyana case, the concession purchased may not include the richest biological communities in Guyana, and the land is in steep, inaccessible terrain where removal of timber would be extremely difficult. Also, there is a strong incentive for the Guyana government to have more land with conservation status in order to meet third-party certification guidelines to gain green-stamp approval for well-managed, environmentally and socially responsible forestry operations and thereby increase export opportunities. The unintended consequence of conservation organizations purchasing low-value forestry concessions for conservation may be an increase in deforestation to export green heartwood, the most popular wood from Guyana, which is found in the more accessible northern part of the country and remains unprotected. Such off-site consequences of land conservation are referred to as "push factors," because protecting one site pushes development to another site. In other words, certain land protection scenarios can result in increases in development or resource extraction in another region, either by removing land supply from production or by providing cash to support development elsewhere. In the case of Guyana, one concession was removed from the market, but the consequences of that action for other concessions that are more threatened remains to be seen. These types of trade-offs are not uncommon in land conservation, and it is important to be aware of them before a full or partial interest in land is purchased.

Because of government corruption and the difficulty of assessing the regional implications of these agreements, a better approach may be to work with local communities that have concessions to develop compensation programs and improve forest management. An example of this approach is a conservation agreement that was made in 2000 between the Wildlands Project and five Mexican conservation groups to protect 2,428 hectares (6,000 acres) of old-growth forest as part of the Sierra Madre Occidental Biological Corridor. The agreement is with the seventy-four-person Ejido Cebadillas, a community cooperative, to protect land that contains approximately half of the nests of the remaining endangered western thick-billed parrot (*Rhynchopsitta pachyrhyncha*) in the world, which was scheduled for logging prior to the agreement. The agreement includes paying the cooperative 50 percent of the net value of the uncut timber over a fifteen-year period. In addition, the conservation organizations agreed to fund a forestry study and sustainable logging plan for the land owned by the cooperative, which will help increase the value of its wood products. The Wildlands Project also plans to help develop ecotourism focused on bird watching. Working with local people will increase the chances of meeting conservation priorities and benefiting local communities.

Restoring Land

Some landscapes have been severely degraded by past land use such that insufficient networks of natural habitat remain and it may be necessary to reestablish habitat. In cases where passive restoration, or what some refer to as rest, may allow native plants to regenerate on their own, that may be the best way to establish habitat for ecological networks (Figure 9.2). In such cases, changes in land management are required. Landowners often will require compensation for activities such as fencing for riparian areas. This type of restoration may require land conservation tools such as the conservation agreements discussed previously. In some areas, rest alone

Figure 9.2. A site along Parson's Creek in northern California with an active restoration treatment (planting trees in grow tubes) in the foreground and passive restoration (fencing riparian area) starting in the middle of the photo along the creek. Both restoration treatments have been established for 2.5 years, but only the passive approach has resulted in increased riparian cover over such a short time period (Opperman and Merenlender 2000). (Photo by Adina Merenlender, 2005.)

may never produce the desired results or may take too long, and more active restoration is required to provide habitat for the planned corridor project. Particular attention needs to be paid to the local habitat types and conditions, as well as desired outcomes. Again, a significant investment should be made in planning a restoration project prior to implementation.

Since establishing corridors, especially at smaller (e.g., single-property) and middle-sized (e.g., watershed) scales, often requires restoration even where remnant forest patches exist, we provide a few suggestions on how to approach ecological restoration projects. In many cases, weeds, including unwanted woody vegetation, may be present in important conservation areas. The removal of unwanted vegetation should always be done in small increments in order to avoid disturbing the site and exposing too much bare soil to the elements, which can result in the erosion of valuable topsoil and colonization of more weeds. Also, small disruptions are less likely to disturb wildlife living in these less desirable communities. In some cases, mature native trees need protection during restoration, especially for their root systems. This can be problematic in highly developed areas. Other important habitat features that should be maintained include naturally wet areas, stream banks and associated riparian areas, and rocky outcrops that provide important habitat for wildlife.

In the case of the country-wide restoration effort that is occurring in South Africa, scientists convinced forward-thinking decision makers that invasive woody species were using more water than native species, leading to reduced runoff and thereby reducing water levels (Le Maitre et al. 1996). Through the hard work of many organizations and individuals, the Working for Water program has grown into a widespread community-based effort to remove invasive plants that is benefiting water conservation, biodiversity, and resulting in much-needed employment (Hobbs 2004).

In some areas, more than just exotic plants need to be removed. Garbage, such as cars, concrete, metal, and other undesirable materials, may need to be removed and perhaps replaced by more natural structural materials such as rocks and logs if it's serving a useful structural purpose. In more contaminated areas, pollution remediation is a necessary first step. If land clearing is required, it is best to begin upslope and clear and replant small, noncontinuous sections of land incrementally over time so that slope stability and existing habitat structure can be maintained. It is always best to establish native vegetation prior to removing adjacent non-native vegetation. If the corridor is being restored for a particular animal species, plantings may need to consider the species' dietary requirements. For example, the Massey Creek project in Queensland, Australia, aims to plant twelve thousand trees to allow cassowaries to move between remnant habitat patches. The trees being planted will include some of the cassowaries' favorite foods (Queensland Parks and Wildlife Service 2003).

In addition to controlling weeds, it is important to discourage invasive vertebrate pests such as bird species that parasitize native species' nests, as well as nonnative rats, foxes, cats, and other exotic predators that consume native species (Burbidge and Manly 2002, Churcher and Lawton 1987). A dramatic and well-documented example of this problem is the introduction in about 1950 of the brown tree snake (*Boiga irregularis*) to the island of Guam (Micronesia), where it led to many local bird species' extinctions (Wiles et al. 2003). This voracious predator caused the extinction of nine out of thirteen native forest bird species of various sizes, at least one of three bats, and at least two species of lizards and stopped three species of pelagic birds from breeding on the island. The snake profited from both a lack of antipredator adaptations in the native vertebrates and the presence of abundant alternative prey in the form of introduced species that had coevolved with predators and helped maintain the snake population (Fritts and Rodda 1998).

Often a combination of passive and active restoration is required. For example, to increase connectivity between Morro do Diablo State Park and adjacent fragments of Atlantic Forest in the state of São Paulo, Brazil, local citizens are working with the Wildlife Trust and Instituto de Pesquisas Ecológicas to protect natural habitat and plant trees. The goal is to provide additional habitat for several endemic species including the black lion tamarin (*Leon topithecus chrysopygus*) (Wildlife Trust). In a similar effort, over 22 hectares (54 acres) of forest was revegetated with native plant species to create the Pinbarren Wildlife Corridor linking ridgetop habitat and riparian areas in Noosa Shire, Queensland. This is a coordinated project involving many landowners who are part of a land care group. They report an increase in bird numbers, echidna sightings, and tortoises since restoring the area (Queensland Parks and Wildlife Service 2003).

Habitat restoration is a field unto itself and if pursued should be researched independently. Useful places to learn more about the science of ecological restoration are two edited volumes, Jordon et al. (1987) and Temperton et al. (2004). For tips on how to do restoration see *The SER International Primer on Ecological Restoration* (Society for Ecological Restoration International Science and Policy Working Group 2004).

Lessons from Corridor Projects

The number of corridor projects worldwide is unknown. In 2001, the International Union for the Conservation of Nature (Bennett and Witt 2001) identified over 150 ecological network projects focused on conserving biodiversity at the landscape or regional scale with an emphasis on ecological interconnectivity, restoring degraded ecosystems, and conserving buffer zones. At the beginning of that effort, researchers developed a useful definition of an ecological network: "a coherent system of natural and/or semi-natural landscape elements that is configured and managed with the objective of maintaining or restoring ecological functions as a means to conserve biodiversity while also providing appropriate opportunities for the sustainable use of natural resources." Thirty-eight networks were surveyed for general information. Few projects had begun an implementation phase, and therefore it was too early to evaluate the effectiveness of these efforts.

The number of projects is certainly larger today and would greatly increase if smaller-scale projects were included. From a search of the World Wide Web, it is astonishing how many wildlife corridor projects seem to be underway, in every country at multiple scales. Clearly, the idea of reconnecting our landscapes has taken hold, and many public and private institutions are involved. The most impressive efforts are being made by local volunteer groups and landowners with relatively few resources, and it can be difficult to access information for those. The selected examples in this chapter include available information from a wide variety of countries and represent different spatial scales, from international networks to road underpasses.

Pan-European Ecological Network

The Pan-European Ecological Network was formed by the Council of Europe to increase habitat connectivity in the near term as one of the primary objectives of the Pan-European Biological and Landscape Diversity Strategy. The Strategy is the European tool for implementing the 1992 Rio Convention, which stated that each nation has a responsibility to maintain its biological diversity. To begin the process, a committee of experts is providing a framework to implement the Network, enabling analysis of existing related programs and encouraging the creation of transnational and transboundary networks to ultimately create a coherent system of biodiversity conservation beyond national boundaries. The focus is on the following components:

- core areas, which are used for the conservation of ecosystems, habitats, species and landscapes;
- biological corridors, intended to improve connections between natural systems;
- restoration areas for the recovery of damaged elements of ecosystems, habitats, and landscapes of European importance; and
- buffer zones, to foster the strengthening of the ecological network and its protection from unfavorable external factors.

Approaches to building ecological networks differ both conceptually and methodologically, according to the differences in project objectives, which are influenced by the nature of human encroachment, social and economic conditions, and natural conditions and scale in various countries and regions (Cook and van Lier 1994).

National-Scale Networks: Moldova's Ecological Network Plan

Each country in Europe has a different approach to the mandate established by the Council of Europe. As an example, we review the Ecological Network plan for the Republic of Moldova. Moldova is a small country (33,371 square kilometers, or 13,015 square miles) northeast of Romania and one of Europe's poorest countries, with approximately 4.2 million people. The landscape is made up of remnant forest and hilly wooded steppe, much of which had been deforested by the end of the nineteenth century for intensive agriculture. Today, much of the land has problems with soil erosion and toxicity as a result of intensive land use and poor agricultural practices—all contributing to continued desertification. Recently, Moldova has embraced several initiatives to protect and restore the environment.

In 2001, BIOTICA Ecological Society, a national nongovernmental organization, developed the Ecological Network of the Republic of Moldova with funding from the republic's National Ecological Fund to integrate a national corridor plan with the Pan-European Ecological Network (Andreev et al. 2002). BIOTICA wants the Eco Network to provide optimal conservation and restoration of biological and landscape diversity, as well as create the prerequisites for sustainable agriculture in Moldova, including a reduction in soil erosion and improvements to the hydrological regime and water quality.

The hope is that if the identified core areas, biological corridors, restoration areas, and buffer zones are protected and restored, the remaining natural resources will be protected and continued desertification will be prevented. The Ecological Network plan identifies areas of high natural value and those that need to be surveyed, as well as sites designated for restoration. The components of the network, which include core areas for conservation and buffer zones, as well as biological corridors at the international, national, and local scales, were mapped at a 1:500,000 scale using GIS. An impressive amount of background information went into this process in order to prioritize components of the network. Spatial data on species and communities, including a revised list of threatened species and endemic communities, as well as protected areas and private land, were used to assess and prioritize sites to include in the network. Core sites were selected primarily according to uniqueness, biodiversity, and vulnerability, and any identified nonforested steppe natural ecosystems were also included. Weighted priorities were used to identify the most important core areas that need to be secured for biological diversity conservation, and those areas are distributed evenly across the country.

The selected international corridors (at least 500 meters—or 1,640 feet—wide) include some core areas and must pass along the entire length of the country to provide habitat connectivity with neighboring nations, as part of the Pan-European Ecological Network. National corridors include core and buffer areas but do not join neighboring countries. Local corridors connect core, buffer, and restoration areas and were primarily established based on hydrologic features, topography, and soils. Existing or projected tree plantations are included in some local corridors. All corridors are accompanied by buffer areas of 50–3,500 hectares (123.5–8,645 acres), with the larger buffers reserved for national corridors. In addition to linear corridors, smaller areas that support Red Data List species are included in the network. Areas in need of restoration due to excessive environmental degradation were also identified.

While a series of useful, well-informed rules were used to designate the network, analysis was not employed to identify where corridors should be established to optimize the desired conservation benefits. A more systematic approach may be employed at the local level, however, using more detailed data. The effort that went into designating this network is impressive, especially given the economic circumstances that Moldova faces and its ongoing struggle with the separatist movement in the Transnistria region. We hope the results of this project will be monitored to determine its effectiveness.

The BIOTICA plan also includes necessary actions required for the implementation of the plan. Some areas will entail changes in land use and active restoration. The plan states that if a change in land use that will restrict economic returns is required to meet the objectives of the Ecological Network, the land in question will be purchased or compensation offered to the private owner. Therefore, implementing the Moldova corridor plan relies on incentive-based conservation, in which the landowners are compensated for financial losses associated with the proposed protections. However, in many places, more than economic value may be at stake when land-use and management changes are enacted. If people live in designated wildlife corridors, changes to the land use can affect their culture in addition to their livelihood, and those changes may have unintended cascading effects that can be hard to predict.

Community-Based Conservation of a Regional Corridor: Kwa Kuchinja Corridor

We have discussed the limitations of parks and the importance of the matrix for species persistence. Nowhere are those two factors more apparent than in Africa, where so many large mammals migrate across a vast landscape. In recent history, most revenues obtained from wildlife hunting have gone to the national governments and therefore have not provided needed assistance at the local level. That is the case in Tanzania, where safari hunting outside of national parks is generally managed by the national wildlife division. As part of the community-based conservation movement, there is a new government policy in Tanzania promoting Wildlife Management Areas (WMA), where local people are supposed to manage wildlife. An important objective of this policy is "to transfer the management of WMA to local communities, thus taking care of corridors, migration routes, and buffer zones; and ensure that the local communities obtain substantial tangible benefits from wildlife conservation" (United Republic of Tanzania 1998). However, the implementation of WMAs has been stymied by regulations and apprehension among local people over land tenure. The concerns are especially strong for the Masai people, who rely on livestock grazing for their livelihood.

Historically, the Masai people who live in this area had a complex range management regime that required them to migrate through areas that received above-average rainfall (Homewood and Rodgers 1991). While Masai grazers still need to take advantage of seasonal water sources, families now live in permanent homesteads across the Tarangire-Manyara Ecosystem (Goldman 2003). Land-use change such as going from extensive to intensive agriculture and the establishment of national parks have made it difficult for Masai pastoralists to manage livestock in this area. For example, land under cultivation in the Kwa Kuchinja corridor has increased from 1,980 hectares (4,891 acres) in 1987 to 6,308 hectares (15,581 acres) in 1998, when the proportion of area used for cultivation was 16.4 percent (see "The KwakuchinjaWildlife Corridor," www.fauna-flora.org/africa/kwakuchinja.html; Figure 9.3). Widespread intensive agriculture also

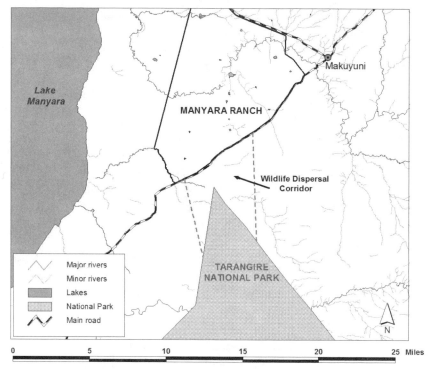

Figure 9.3. The Kwa Kuchinja wildlife dispersal corridor between Tarangire National Park and Manyara Ranch in Tanzania. (Permission to use by Lara Foley, Wildlife Conservation Society and African Wildlife Foundation.)

excludes large mammals. International conservation nongovernmental organizations have begun to try to protect the Kwa Kuchinja wildlife corridor and the Masai culture that depends on conserving undeveloped open space.

One local village in the corridor has expressed interest in setting aside part of its land to enhance wildlife movement in return for economic compensation. Rather than working through the government WMA program, some villagers are working with the new Tanzania Land Conservation Trust and Fauna and Flora International to identify a mechanism for generating revenue at the local level for the land set aside. In addition to these activities, the Tanzania Land Conservation Trust has recently purchased the Manyara Ranch (17,800 hectares, or 43,966 acres) to augment the protection of the Kwa Kuchinja wildlife corridor (http://www.awf.org/success/manyara.php).

So far, implementing the Kwa Kuchinja wildlife corridor has involved land conservation tools such as purchasing land and trying to maintain grazing land that, it is hoped, can still provide a livelihood and important cultural resources for the Masai people, as well as necessary open space for wildlife migration. While incentive-based conservation is touted by some as the solution to local control over resources, there remains a need to absorb local knowledge about pastoral land management before a win-win solution is likely to develop. Community-based conservation needs to go beyond extending information to local people and should aim to involve them in land-use planning (Goldman 2003).

Large-Scale Corridor Planning in Urban Areas: Lisbon's Master Plan

We wrote in Chapter 4 that landscape features, such as riparian areas, often serve as de facto ecological connectivity. In urban areas, rivers and streams historically served as transportation corridors, and more recently they are the focus of urban renewal projects, because people enjoy walking and recreating along scenic waterways. A large-scale planning effort is underway in the greater metropolitan area of Lisbon, Portugal (3,100 square kilometers, or 1,209 square miles), to enhance the environmental, flood management, and aesthetic properties of river corridors running north-south through deeply incised valleys to the coast (Saraiva et al. 2002). This region supports the highest population density in Portugal (2.5 million people in nineteen municipalities); hence, the existing built environment greatly constrains restoration options. Rapid development in the area and a lack of environmental planning have led to problems with water quality from sewage, loss of stream function and habitat from concrete-lined channels, and other in-stream alterations, resulting in increased flooding downstream.

Some of the changes to the stream channels and floodplains were mapped from historic aerial photography, and current channel and riparian condition was assessed in the field. Stream attributes such as bank type, riparian structure and species diversity, and apparent water quality were collected. An impressive amount of work went into surveying stream cross sections to determine existing bank and floodplain characteristics and conditions. This information along with information about constraints from the built environment was used to identify reaches with potential for restoration or conservation (Figure 9.4).

In addition to the effort that was put into assessing a vast hydrologic network, this corridor project, which involves four stream corridors, has

Figure 9.4. The condition of one stream corridor site near Lisbon, Portugal, as it currently exists (above) and what is being proposed for the same site (below). (Permission to use by Graça Saraiva.)

one of the most comprehensive implementation plans in place for a highly urbanized region. Coordination among the necessary partners, including regional and local authorities, is outlined, as well as sponsorship by the company responsible for several municipal sewage systems. The plan addresses several overarching improvements to urban development, transportation, social infrastructure, sewage systems, environmental protection, and nature conservation.

This plan is helping regional interest groups develop a new approach to river management and flood protections based on entire river catchments and restoration of stream processes, rather than the more traditional engineered solutions used in the past. It is hoped that implementing the plan will allow for increased water infiltration and provide additional water recharge. In addition to enhancing the hydrologic connectivity and restoring riparian communities, this plan includes many efforts to improve connectivity and enjoyment for people who live and work in the area, primarily by enhancing the scenery and recreation opportunities through the creation of streamside trails (Figure 9.4).

This effort, spearheaded by landscape architects, demonstrates the different outcomes that can result, depending on the goals of the project. In this case, the focus is on improving aesthetic elements of the landscape while considering the river continuum concept (Vannote et al. 1980). Hence, designs for individual streamside parks associated with urban renewal are the starting point rather than connectivity per se. Certain conservation benefits are likely to result from this approach, such as increased riparian cover and improved water quality, but some native species may not be compatible with the human use at the sites. Nevertheless, this project is an excellent example of how connectivity can be enhanced for human systems and address urban ecology issues.

Maintaining Connectivity within a Watershed: Sonoma Valley and Mountains to Mangroves

Conserving watershed process and function has been a focus for many people living in Sonoma Valley, California. By definition, a watershed is the land that water flows across, or under, on its way to a stream, river, lake, or ocean—in other words, the area drained by a river or river system. Watershed planning has been embraced by local community groups as a way of improving the local environment. These watershed groups have become a popular means of collaboration and community-based land stewardship (Kennedy 2000). This type of systems approach represents an important mechanism to integrate environmental, public, and economic welfare.

The Sonoma Creek watershed drains into the northern part of the San Francisco Bay and is under intense pressure from increasing population and changing economies, resulting in increased habitat fragmentation. Most of the lowland habitats, such as marshes, gently sloped streams, and oak savannas, are gone, replaced with buildings, roads, and agriculture. Woodlands and forestlands were converted to vineyard, and those agricultural areas along with the required fencing can present a barrier to animal movement (Hilty and Merenlender 2004, Merenlender 2000). Most of the protected natural areas are on the surrounding hillsides, and very little habitat remains in the valley connecting those areas.

To increase connectivity in this watershed, the Sonoma Ecology Center, a nonprofit watershed conservation group, is working to protect one of the last swaths of natural habitat that crosses the Sonoma valley floor as a wildlife habitat corridor. The proposed corridor is about 8 kilometers (5 miles) long and up to 1.6 kilometers (1 mile) wide, encompasses most of the region's habitat types, and connects isolated high-elevation natural areas to a major stream (Figure 9.5).

Many of the properties within the proposed corridor have been inventoried to map existing fences, habitat, and wildlife movement, and likely barriers to animal movement have been identified. The most obvious one is a two-lane highway that divides the corridor. Wildlife surveys using remotely triggered cameras demonstrated that the undersized culverts fail to facilitate animal passage under the highway. The culverts were due to be widened, but the California budget shortfall has delayed the improvements. Other obstacles within the proposed corridor include fences that prevent animal movements and eroding fire roads.

The land is in both public and large private ownerships, reducing the number of negotiations required to protect the site. Even a project like this, however, which appears at the onset to be highly feasible, requires a

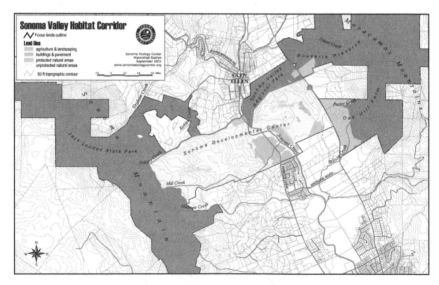

Figure 9.5. The proposed Sonoma Valley, California, habitat corridor connects two higher-elevation protected areas (dark gray) and primarily consists of public land managed by the Sonoma Development Center (lighter gray with black outline) that transects urban development along Highway 12. (Permission to use figure by Sonoma Ecology Center.)

tremendous amount of effort and coordination. Private owners of large or significant parcels within and adjacent to the proposed corridor have been interviewed about their views on wildlife, fences, and future habitat enhancement on their property. The Sonoma Ecology Center is trying to facilitate the removal or alteration of the fences most problematic for wildlife movement and enhance habitat by increasing appropriate native vegetation. The center is working with willing landowners and local funders to try to secure long-term commitments from the primary landowners and agencies to manage the proposed corridor in perpetuity for wildlife and native plant communities. After the site is protected, the priorities will shift to management and restoration to maximize connectivity.

This project demonstrates some of the design, planning, and implementation issues that we discuss in this book. Early surveys and analyses have documented the use of the proposed corridor by focal species and the potential barriers that need to be addressed. A geographic information system has been essential for design, planning, and communicating the project to the public. The Sonoma Ecology Center has long-standing positive relationships with multiple interest groups and institutions in the region and is working with local landowners. Through this project, the importance of habitat connectivity has become better understood by the general public and by decision makers. The Sonoma Ecology Center has increased its understanding of the natural and human systems that make up the Sonoma Valley habitat corridor, which is going to greatly increase its chances of successfully enhancing connectivity across the watershed. Unfortunately, the adjustments that need to be made to protect this one corridor are expensive and require participation from the state transportation department as well as private landowners. Success depends on raising support and awareness of the issue beyond the local community. Maintaining other corridors across the valley would help the problem of having all the eggs in one basket.

Another community project focused on reconnecting natural areas from the top to the bottom of a watershed is the Mountains to Mangroves project, which started in 1995 just north of Brisbane, Australia's third-largest city, with a population of 1.6 million people. The Mountain to Mangroves Committee states the following on its Web site: "This bush land corridor links us to our cultural heritage and connects us to our wild neighbors" (http://www.mountainstomangroves.org).

This corridor was explicitly designed to diminish the impacts of ongoing forest fragmentation by reconnecting remnant forest patches to provide a wildlife corridor. The Mountains to Mangroves Committee Foundation, which includes representatives from interest groups, local and state government, and state parks, worked with consultants to develop a strategic plan. The corridor plan includes enhancing both environmental values and recreation. In addition to setting out objectives for the project, details on land management and tasks that need to be accomplished to protect the wildlife corridor are outlined. The Mountains to Mangroves Corridor Committee acts to protect habitat connectivity in the watershed by opposing new barriers, such as road extensions, and promoting sustainable development. It has also been active in lobbying to protect state-owned land from development in order to retain natural habitat.

This well-coordinated local effort has brought important human and financial resources to the protection and enhancement of watershed connectivity. Local community members identify with the Mountains to Mangroves (M2M) corridor and come together to celebrate each year at the M2M Festival. These types of activities and the extension of information that the project provides have increased important connections between people and place. However, according to the project extension officer, more work needs to be done to extend information on the importance of the corridor for enhancing connectivity for the endemic fauna and flora. In this case, the community has an increased appreciation for the place where they live, but that has not necessarily translated into a more wildlife-friendly landscape, in part due to increased development.

Removing Barriers at the Site Scale: Highway 93, Montana

As we discussed earlier in the book, roads can present a serious barrier for animal movement. Some of the most common work that is being done to reconnect our landscapes is to mitigate the impacts of roads on wildlife because of the significant threat that they pose. An excellent book on roads and the problems posed for wildlife conservation, as well as ways to improve the situation, is *Road Ecology: Science and Solutions* (Forman et al. 2003). Mitigating the impacts of roads is primarily done by public transportation departments. However, some local interest groups have become involved in investigating alternative road design and impacts of roads on local wildlife. In Montana, the Confederated Salish and Kootenai Tribes, the Montana Department of Transportation, and other groups recognized an opportunity to protect wildlife when proposals arose for the reconstruction of U.S. Highway 93, which passes through the Flathead Indian Reservation (Figure 9.6). Grizzly bear (*Ursus arctos*), wolverine (*Gulo gulo*), white-tailed (Odocoileus virginianus) and mule deer (Odocoileus hemionus), pronghorn antelope (Antilocapra americana), elk (Cervus elaphus), painted turtles (Chrysemys picta), bighorn sheep (Ovis canadensis), and numerous other species survive on the reservation and cross the highway, too often unsuccessfully. Modeling also indicated that part of the reservation may be a key linkage area for wildlife connectivity between the Selway-Bitterroot wilderness area in Idaho and the Mission Mountain–Bob Marshall wilderness area in Montana (Mietz 1994). In discussions about the potential highway reconstruction, the tribes stressed the importance of maintaining the integrity of their land and their culture, including their wildlife heritage. Maintaining and restoring natural processes, including the movement of wildlife, became a major focus of this project, and landscape architects and wildlife biologists developed a workbook on the subject (Jones & Jones Architects and Landscape Architects, Ltd. 2000).

Ultimately, the installation of wildlife-proof fencing perforated with animal crossing structures was recommended in the "Design Guidelines and Recommendations" section of the workbook (p. 25). The fencing and structures were modeled after those on the Trans-Canada Highway in Banff National Park in Alberta (Figure 7.3), and the recommended placement of the structures depended on local knowledge and data, including game trail and road kill locations (Figure 9.6), preferred habitat types, and engineering feasibility. In 2000, the tribes, the Federal Highway Administration, and the Montana Department of Transportation signed a Memorandum of Agreement to include 24 kilometers (15 miles) of fencing and forty-two crossing structures with 74 kilometers (46 miles) of road reconstruction from Evaro to Polson (except for a 16-kilometer, or 10-mile, section that is still under evaluation). Arguably, this highway project represents the most extensive effort to accommodate safe wildlife crossing in the United States. The cost of the project is estimated to be over US$9 million.

Another positive aspect of this project is the commitment to support ecological monitoring to evaluate the effectiveness of the wildlife crossings and fencing by the Western Transportation Institute and associated graduate students. The information obtained from monitoring the improvements on U.S. 93 will be important for managing wildlife on the reservation and pivotal to future decisions nationwide regarding wildlife crossing structures and fencing installations.

This corridor project is a wonderful example of the advantages of collaborating with scientists in project planning and implementation, an ap-

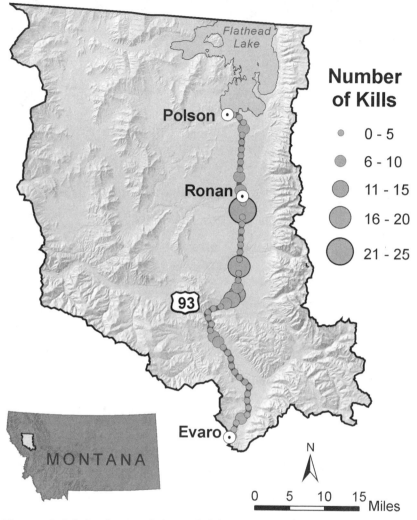

Number of Kills

- 0 - 5
- 6 - 10
- 11 - 15
- 16 - 20
- 21 - 25

Flathead Lake

Polson

Ronan

93

Evaro

MONTANA

N

0   5   10   15 Miles

Figure 9.6. Relative density of white-tailed deer (*Odocoileus virginianus*) and mule deer (*O. hemoinus*) killed along Highway 93 between 1998 and 2003. The data are from the Montana Department of Transportation (MDT) and the Montana Highway Patrol and were recorded as the number of individual road kills observed per 0.16 kilometer (0.1 mile). Data were compiled by Whisper Camel, a student at Montana State University, and have not been verified by Montana Department of Transportation as to accuracy. The discussion, opinions, and conclusions presented in this book are not necessarily those of MDT; design alternatives to address this issue are experimental in nature and are still being finalized.

proach we advocate in Chapter 8. Together the Confederated Salish and Kootenai Tribes, public transportation organizations, and scientists have set a precedent for future road reconstruction projects to conserve connectivity.

In Summary

These are just a few of the examples of corridor projects that are going on around the world at multiple scales, from highway underpasses to Pan-European networks, involving individual landowners, as well as local, national, and international organizations. We highly recommend that those interested in enhancing and restoring connectivity visit and learn about other projects in order to take advantage of lessons learned. The most successful projects have strong collaborative relationships at all scales of governance and across interest groups. Building those relationships is essential to any planning process and will increase understanding of the human and natural systems.

Conclusion

The potential benefits of as well as the problems associated with maintaining and restoring connectivity depend a great deal on environmental, social, and historical context. Therefore, throughout this book, we used many different examples from the scientific and popular literature to illustrate (1) the various reasons for and consequences of habitat loss and fragmentation; (2) how ecological principles apply to species and communities; (3) different types, purposes, and efficacies of corridors; and (4) project planning and implementation. By exploring the problem from a theoretical perspective and presenting practical examples, we aimed to make the information in this book useful to those studying and practicing conservation science in a wide variety of situations. Because context is so important, no one size fits all with respect to corridors. However, where possible, we provided ecological principles to consider and general steps that may improve the chances of successfully maintaining and restoring connectivity.

Scale is the most defining aspect of a project and is influenced by the nature of the problems addressed, the extent of existing barriers to connectivity, conservation objectives, and the institutions addressing the problem. The project scale, in turn, influences the processes that can be maintained or restored. For example, large-scale connectivity may be required to facilitate large-animal migration, whereas plant pollination may be accommodated within a smaller protected area network. We illustrated this range of needs by describing small-scale projects, from that of the school class in Australia that restored a highway underpass for wildlife movement on up to the efforts of defining ecological networks for all of

Europe. Institutional participation greatly influences the scale at which a problem is addressed, since locally based environmental groups rarely take on large-scale planning efforts, and political entities find it difficult to work beyond their boundaries.

No matter the scale of an existing project, it is important to at least consider the larger landscape context in planning, design, and evaluation of ecological corridors. This is because large-scale processes often influence local-scale phenomena. Also, as time passes, there are often new efforts to address connectivity at larger scales that will need to build on existing corridors and reserves. It is also important not to rely on the minimum dimensions or on a limited number of landscape features to provide connectivity, because a variety of larger features will be more resilient to change over time. This is particularly true given the large number of problems that can exist when relying solely on corridors for biodiversity conservation. These include genetic and demographic problems that can plague species in marginal and fragmented habitat. Edge effects can sometimes permeate an entire ecological network, and impacts to species conservation can include predation and competition from exotic species and other processes that may result in low survivorship for native species.

We know that animal and plant populations vary from being completely isolated from one another, through intermittently connected metapopulations, to sharing a single widespread breeding population. Some populations require different contiguous ecosystems, while others require two or more habitats that may be thousands of kilometers from one another. Therefore, conserving and enhancing connectivity usually requires more than a single, minimum-size corridor. Ecological networks need to be as resilient as possible to perturbation, even from large-scale stochastic events. In principle, habitat units will ideally be large enough to withstand expected perturbations or catastrophic events that may affect connectivity. If not, they will require extensive and costly management and ultimately may fail to provide adequate protection for biodiversity.

Moreover, we need to acknowledge that land-use planning must realistically be based on less than adequate information, reducing our ability to make accurate predictions regarding the response of species and natural communities to proposed changes across the landscape. Lack of funding may make it difficult to augment existing data and provide for monitoring the impacts of conservation strategies as they proceed. Even more difficult is the reality that planning, more often than not, proceeds in a social context that constrains planning options and hence biases the process toward minimalist rather than optimalist solutions. Given these constraints, land-use planning should adopt an adaptive management strategy to the extent possible, to allow for adjustments in the process as knowledge and experience are gained.

Understanding the historical context of a landscape is important. Geologic history plays a critical role in shaping natural ecosystems; and, more recently, human history has determined the extent, type, and complexity of the matrix. Conservation biologists are increasingly recognizing the importance of the matrix on biodiversity both in and outside of reserves. The matrix has particular influence on ecological networks because they often incorporate narrow, marginal habitat surrounded by modified landscapes. The type and configuration of the matrix influences species invasions and impacts, and therefore influences which native species are likely to persist.

Mapping the history of land-use change is often the first step to developing models used to forecast future land use. While there are many uncertainties surrounding forecasting models, it is important that we do our best to make predictions of where existing natural areas are in need of protection to prevent future habitat loss and fragmentation. Protecting natural resources and connectivity that are threatened in the near term is generally more costly than protecting landscapes that are not under demand for development. However, there are often critical environmental benefits that may be lost due to land-use change in the immediate future; therefore, resources that are under immediate threat should be a priority for conservation. Land use continues to fragment and impact natural systems, as we discussed, but land-use change does not receive enough attention as a primary determinant of global change. The problems associated with continued land use that we do not address today will simply be passed on to future generations, providing them fewer options for conservation and imparting higher costs.

Future Goals and Directions

We lack a great deal of information on ecological communities, landscape-level processes, and interactions between natural and human systems. This is particularly true in certain parts of the world. South America, Africa, and southern Asia in particular have been studied relatively little, especially in view of the biological complexity that we know occurs in those vast regions. Even in better-studied areas, more research is needed in conservation science, landscape ecology, and restoration ecology. In fact, con-

servation biology is such a new discipline that we can confidently anticipate many new insights and even surprises to be forthcoming.

Scientists often document changes in natural communities caused by
human impacts, but in most cases we lack an adequate understanding of
the underlying mechanisms driving the patterns of change we have documented. Besides habitat destruction and introduction of exotic species,
humans undoubtedly affect wildlife activities and well-being in many
other ways that have been little studied. The interface of human-dominated and other habitats needs continual exploration. For example, the
widespread phenomenon of exurban development needs better mapping,
monitoring, and modeling. This will require a deeper understanding of
the development process and impacts to natural systems. One of the consequences of development is the spread of invasive species. The relationship between landscape configuration, including the pattern and type of
the matrix, and the spread of invasive species needs to be explored in various bioregions.

We are beginning to understand how natural systems work at landscape
scales, but many refinements are needed, and we are far from being able
to predict landscape behavior in specific situations. Moreover, more attention needs to be paid to spatial ecosystem modeling that integrates biological and physical (and if possible social) processes across large-scale
ecosystems and entire regions. Environmental modeling (e.g., hydrologic
and air pollution) needs to be combined with land-use change modeling
to forecast the environmental consequences of likely future land-use
change. This is just one example of where model integration across disciplines will be most useful to conservation planners and decision makers.
These types of models and the increased scientific understanding and practical applications that they can provide will become even more critical in
a prolonged era of rapid climate change.

Presently, we know relatively little about how organisms move through
landscapes, the cues they use to navigate, and the resources that are critical for their safe passage. These are fundamental questions for corridor design. Our understanding of how species respond to ecotones of various
types is also in its infancy. The interesting question of whether edge zones
are stable over time or will gradually grow in width, further diminishing
the effective size of habitat patches, needs to be addressed.

We have much to learn about the nature of interactions among species
in communities, and we must strive to more accurately predict how those
interactions will change if a species disappears or conversely an exotic

species invades. In general, much more needs to be learned about the behavior of small populations and the generation of minimum threshold densities. In addition, the roles of parasites and diseases, both native and exotic, have been especially poorly investigated.

In our efforts to reduce the largest extinction episode in Earth's history, we will need all the tools in our toolbox and more. In addition to the obvious first steps, such as trying to maintain minimum viable populations of all native species, we need to protect the earth's remaining wilderness areas and the goods and services they provide. To do that, we need to better quantify the relationship between biodiversity and those services and better allocate resources for land conservation. Identifying the location and extent of protected habitat that will best achieve multiple project objectives or conservation benefits requires that we quantify the functional relationship between conserving land and the results, such as the persistence of a focal species. This can be difficult when species occurrence and demographic data are limited and require quantifying the combined value of multiple benefits, from species conservation to water resources and recreational amenities—all common objectives of land conservation. Many of the benefits are not fungible and therefore cannot be interchanged, and, in some cases, one location may not be substitutable for another. Research is needed to examine the best methods for valuing combinations of conservation benefits to more effectively target land conservation and enhance connectivity for multiple objectives.

A major challenge of all land-planning efforts is the recognition that all places are idiosyncratic; that is, planning must be site based. The best available science and extensive experience are important, but they are only the starting points for knowing and understanding local conditions. It is that local context that ultimately must direct planning if it is to be successful. Moreover, it is increasingly apparent that more and more resources will need to be applied to reversing the current trend of deterioration. This need confronts the paradox that while it will cost us more for conservation and restoration, the levels of poverty in the world are rising rapidly, as is the concomitant level of expectations for improved quality of life for the world's human inhabitants.

For this scientific enterprise to be useful to conservation, it must be continually translated and disseminated for applied purposes and be incorporated into education and outreach at all levels. A more concerted effort needs to be made to educate young people about the natural world and our connection to it. Practitioners working for government, private, and nongovernmental organizations must be well informed and receive needed support in order to implement potential solutions. Unfortunately, resources are almost never made available for disseminating information, despite the desire of research funding organizations to ensure that the science they support has an impact. Conservation science and conservation practice are inextricably linked, and it is well recognized that ecosystem management is an adaptive process that requires continuous monitoring. We have emphasized throughout this book the importance of ecological monitoring for effective conservation planning and project implementation.

Finally, we applaud the growing community of scientists, environmental activists, planners, land managers, and politicians who are working on behalf of the world's citizenry to prevent further fragmentation, to restore connectivity at all scales, and in the end to make the world healthier for all of its inhabitants. It is our hope that this book will prove to be useful to those students, researchers, practitioners, and policy makers engaged in the study and implementation of ecological networks.

References

Abbitt, R., J. Scott, and D. Wilcove. 2000. The geography of vulnerability: Incorporating species geography and human development patterns into conservation planning. *Biological Conservation* 96:169–175.

Åberg, J., G. Jansson, J. E. Swenson, and P. Angelstam. 1995. The effect of matrix on the occurrence of hazel grouse (*Bonasa bonasia*) in isolated habitat fragments. *Oecologia* 103:265–269.

Abson, R. N. 2004. The use by vertebrate fauna of Slaty Creek Wildlife Underpass, Calder Freeway, Black Forest, Macedon, Victoria. Master's thesis, University of Tasmania, Australia.

Adler, G. H., and R. Levins. 1994. The island syndrome in rodent populations. *Quarterly Review of Biology* 69:473–490.

Agarwal, C. G., J. Green, T. Grove, T. P. Evans, and C. M. Schweik. 2002. *A review and assessment of land-use change models: Dynamics of space, time and human choice*. General Technical Report NE-297. Newton Square, PA: U.S. Department of Agriculture, Forest Service, Northeastern Research Station.

Ahern, J. 1995. Greenways as a planning strategy. *Landscape and Urban Planning* 33:131–155.

Albrecht, T. 2004. Edge effect in wetland-arable land boundary determines nesting success of scarlet rosefinches (*Carpodacus erythrinus*) in the Czech Republic. *Auk* 121:361–371.

Allee, W. C. 1931. *Animal aggregations: A study in general sociology*. Chicago: University of Chicago Press.

Allee, W. C. 1951. *Cooperation among animals, with human implications*. New York: Henry Schuman.

Allen, J. C., W. M. Schaffer, and D. Rosko. 1993. Chaos reduces species extinction by amplifying local population noise. *Nature* 364:229–232.

Anderson, P. K. 1970. Ecological structure and gene flow in small mammals. *Symposium Zoological Society of London* 26:299–325.

Ando, A., J. Camm, S. Polasky, and A. Solow. 1988. Species distributions, land values, and efficient conservation. *Science* 279:2126–2128.

Andreassen, H. P., S. Halle, and R. A. Ims. 1996. Optimal width of movement corridors for root voles: Not too narrow and not too wide. *Journal of Applied Ecology* 33:63–70.

Andreassen, H. P., K. Hertzberg, and R. A. Ims. 1998. Space use responses to habitat fragmentation and connectivity in the root vole *Microtus oeconomus*. *Ecology* 79:1223–1235.

Andreasson, H. P., and R. A. Ims. 2001. Dispersal in patchy vole populations: Role of patch configuration, density dependence, and demography. *Ecology* 82:2911–2926.

Andreassen, H. P., R. A. Ims, and O. K. Steinset. 1996. Discontinuous habitat corridors: Effects on male root vole movements. *Journal of Applied Ecology* 33:555–560.

Andreev, A., P. Gorbunenko, O. Cazanteva, A. Munteanu, A. Negru, I. Trombitki, M. Coca, and G. Sîrodoev. 2002. *Concept of National Ecological Network of the Republic of Moldova*. Chisinau: BIOTICA Ecological Society and Regional Ecological Centre of Moldova. http://www.biotica-moldova.org/Eco-Net/.

Andrén, H. 1992. Corvid density and nest predation in relation to forest fragmentation: A landscape perspective. *Ecology* 73:794–804.

Andrén, H. 1994. Effects of fragmentation on birds and mammals in landscapes with different proportions of suitable habitat: A review. *Oikos* 71:355–366.

Aquilani, S. M., and J. S. Brewer. 2004. Area and edge effects on forest songbirds in a non-agricultural upland landscape in northern Mississippi, USA. *Natural Areas Journal* 24:326–335.

Araiza, A. E. 2005. Arizona road kill: Huge toll on park-area highways. *Arizona Daily Star*, May 16.

Arango-Velez, N., and G. H. Kattan. 1997. Effects of forest fragmentation on experimental nest predation in Andean cloud forest. *Biological Conservation* 81:137–143.

Araújo, M. B., and P. H. Williams. 2000. Selecting areas for species persistence using occurrence data. *Biological Conservation* 96:331–345.

Arendt, R. 2004. Linked landscapes creating greenway corridors through conservation subdivision design strategies in the northeastern and central United States. *Landscape and Urban Planning* 68:241–269.

Associated Press. 1998. Scientists scratch their heads over missing lynx population. November 15, http://lynx.uio.no/lynx/nancy/news/ny.top.htm.

Azmier, J., and L. Stone. 2003. *The rural west: Diversity and dilemma*. Calgary: Canada West Foundation. http:/www.cwf.ca.

Baker, H. G. 1965. Characteristics and modes of origin of weeds. In *The genetics of colonizing species*, ed. H. G. Baker and G. L. Stebbins, 147–172. New York: Academic Press.

Bani, L., M. Baietto, L. Bottoni, and R. Massa. 2002. The use of focal species in designing a habitat network for a lowland area of Lombardy, Italy. *Conservation Biology* 16:826–831.

Banks, S. A., and G. A. Skilleter. 2005. Intertidal habitat conservation: Identifying conservation targets in the absence of detailed biological information. *Aquatic Conservation: Marine and Freshwater Ecosystems* 15:271–288.

Banks, S. C., D. B. Lindenmayer, S. J. Ward, and A. C. Taylor. 2005. The effect of habitat fragmentation via forestry plantation establishment on spatial genotype structure in the small marsupial carnivore *Antechinus agilis*. *Molecular Ecology* 14:1667–1670.

Baranga, J. 1991. Kibale forest game corridor: Man or wildlife? In *Nature conservation 2: The role of corridors*, ed. D. A. Saunders and R. J. Hobbs, 371–376. Chipping Norton, New South Wales, Australia: Surrey Beatty & Sons.

Barbour, M., and J. Major, eds. 1988. *Terrestrial vegetation of California*. Special Publication No. 9. Sacramento: California Native Plant Society.

Barrett, G. W., J. D. Peles, and S. J. Harper. 1995. Reflections on the use of experimental landscapes in mammalian ecology. In *Landscape approaches in mammalian ecology and conservation*, ed. W. Z. Lidicker Jr., 157–174. Minneapolis: University of Minnesota Press.

Batáry, P., and A. Báldi. 2004. Evidence of an edge effect on avian nest success. *Conservation Biology* 18:389–400.

Bawa, K. S. 1990. Plant-pollinator interactions in tropical rain forests. *Annual Review of Ecology and Systematics* 21:399–422.

Beardsley, K., and D. Stoms. 1993. Compiling a digital map of areas managed for biodiversity in California. *Natural Areas Journal* 13:177–190.

Beazley, K., and N. Cardinal. 2004. A systematic approach for selecting focal species for conservation in the forests of Nova Scotia and Maine. *Environmental Conservation* 31:91–101.

Behera, M. D., C. Jeganathan, S. Srivastava, S. P. S. Kushwaha, and P. S. Roy. 2000. Utility of GPS in classification accuracy assessment. *Current Science* 79:1696–1700.

Beier, P. 1993. Determining minimum habitat areas and habitat corridors for cougars. *Conservation Biology* 7:94–108.

Beier, P. 1995. Dispersal of juvenile cougars in fragmented habitat. *Journal of Wildlife Management* 59:228–237.

Beier, P., and S. Loe. 1992. A checklist for evaluating impacts to wildlife movement corridors. *Wildlife Society Bulletin* 20:434–440.

Beier, P., and R. F. Noss. 1998. Do habitat corridors provide connectivity? *Conservation Biology* 12:1241–1252.

Beier, P., K. L. Penrod, C. Luke, W. D. Spencer, and C. Cabanero. 2006. South coast missing linkages: Restoring connectivity to wildlands in the largest metropolitan area in the United States. In *Connectivity and conservation*, eds. K. R. Crooks and M. A. Sanyayan, in press. Cambridge: Cambridge University Press.

Beier, P., M. Van Dreilen, and B. O. Kankam. 2002. Avifaunal collapse in West African forest fragments. *Conservation Biology* 16:1097–1111.

Beissinger, S. R., and D. R. McCullough, eds. 2002. *Population viability analysis*. Chicago: University of Chicago Press.

Beissinger, S. R., and M. I. Westphal. 1998. On the use of demographic models of population viability in endangered species management. *Journal of Wildlife Management* 62:821–841.

Ben-David, M. T., A. Handley, and D. M. Schell. 1998. Fertilization of terrestrial vegetation by spawning salmon: The role of flooding and predator activity. *Oikos* 83:47–55.

Bennett, A. F. 1990. Habitat corridors and the conservation of small mammals in a fragmented forest. *Landscape Ecology* 4:109–122.

Bennett, A. F. 1991. Roads, roadsides and wildlife conservation: A review. In *Nature conservation 2: The role of corridors*, eds. D. A. Saunders and R. J. Hobbs, pages 99–118. Chipping Norton, New South Wales, Australia: Surrey Beatty & Sons.

Bennett, A. F. 1999. *Linkages in the landscape: The role of corridors and connectivity in wildlife conservation*. Gland, Switzerland: World Conservation Union.

Bennett, A. F., K. Henein, and G. Merriam. 1994. Corridor use and the elements of corridor quality: Chipmunks and fencerows in a farmland mosaic. *Biological Conservation* 68:155–165.

Bennett, G., and P. Witt. 2001. *The development and application of ecological networks: A review of proposals, plans and programmes*. IUCN Report B1142. Gland, Switzerland: World Conservation Union and AIDEnvironment.

Bentley, J. M., C. P. Catterall, and G. C. Smith. 2000. Effects of fragmentation of araucarian vine forest on small mammal communities. *Conservation Biology* 14:1075–1087.

Berger, J. 2004. The last mile: How to sustain long-distance migration in mammals. *Conservation Biology* 18:320–331.

Berling-Wolf, S., and J. Wu. 2004. Modeling urban landscape dynamics: A review. *Ecological Research* 19:119–129.

Best, A. 2005. How dense can we be? *High Country News*, June 13: 8–15.

Bian, L. 2003. The representation of the environment in the context of individual-based modeling. *Ecological Modelling* 159:279–296.

Bierregaard, R. O. J., W. F. Laurance, J. W. J. Sites, A. J. Lynam, R. K. Didham, M. Andersen, C. Gascon et al. 1997. Key priorities for the study of fragmented tropical ecosystems. In *Tropical forest remnants: Ecology, management, and conservation of fragmented communities*, eds. W. F. Laurance and R. O. Bierregaard, pages 515–525. Chicago: University of Chicago Press.

Bierregaard, R. O., T. E. Lovejoy, V. Kapos, A. A. Dossantos, and R. W. Hutchings. 1992. The biological dynamics of tropical rain-forest fragments. *BioScience* 42:859–866.

Bird, B. L., L. C. Branch, and D. L. Miller. 2004. Effects of coastal lighting on foraging behavior of beach mice. *Conservation Biology* 18:1435–1439.

Bissonette, J. A., and S. Broekhuizen. 1995. *Martes* populations as indicators of habitat spatial patterns: The need for a multiscale approach. In *Landscape approaches in mammalian ecology and conservation*, ed. W. Z. Lidicker Jr., pages 95–121. Minneapolis: University of Minnesota Press.

Bolger, D. T., T. A. Scott, and J. T. Rotenberry. 1997. Breeding bird abundance in an urbanizing landscape in coastal Southern California. *Conservation Biology* 11:406–421.

Bolger, D. T., T. A. Scott, and J. T. Rotenberry. 2001. Use of corridor-like landscape structures by bird and small mammal species. *Biological Conservation* 102:213–224.

Bollinger, E. K., and T. A. Gavin. 2004. Responses of nesting bobolinks (*Dolichonyx oryzivorus*) to habitat edges. *Auk* 121:767–776.

Boocock, C. N. 2002. *Environmental impacts of foreign direct investment in the mining sector in sub-Saharan Africa*. CCNM Global Forum on International Investment, OECD Conference on Foreign Direct Investment and the Environment: Lessons to Be Learned from the Mining Sector. Paris: Organization for Economic Cooperation and Development.

Boose, A. B., and J. S. Holt.1999. Environmental effects on asexual reproduction in *Arundo donax*. *Weed Research* 39:117–127.

Borgmann, K. L., and A. D. Rodewald. 2004. Nest predation in an urbanizing landscape: The role of exotic shrubs. *Ecological Applications* 14:1757–1765.

Bowers, M. A., and J. L. Dooley Jr. 1999. A controlled, hierarchical study of habitat fragmentation: Responses at the individual, patch, and landscape scale. *Landscape Ecology* 14:381–389.

Bradly, K. 1991. A data bank is never enough: The local approach to landcare. In *Nature conservation 2: The role of corridors*, eds. D. A. Saunders and R. J. Hobbs, pages 377–386. Chipping Norton, New South Wales, Australia: Surrey Beatty & Sons.

Bright, P. W., and P. A. Morris. 1996. Why are dormice rare? A case study in conservation biology. *Mammal Review* 26:157–187.

Brooker, L., M. Brooker, and P. Cale. 2001. Animal dispersal in fragmented habitat: Measuring habitat connectivity, corridor use, and dispersal mortality. *Conservation Ecology* 3:4. http://www.consecol.org/.

Brooks, C. N., and A. M. Merenlender. 2001. Determining the pattern of oak woodland regeneration for a cleared watershed in northwest California: A necessary first step for restoration. *Restoration Ecology* 9:1–12.

Brosset, A., P. Charles-Dominique, A. Cockle, J. F. Cosson, and D. Masson. 1996. Bat communities and deforestation in French Guiana. *Canadian Journal of Zoology* 74:1974–1982.

Brown, J. H. 1971. Mammals on mountain tops: Nonequilibrium insular biogeography. *American Naturalist* 105:467–478.

Brown, J. H. 1978. The theory of insular biogeography and the distribution of boreal birds and mammals. *Great Basin Naturalist Memoirs*, no. 2:209–227.

Brown, J. H., and M. V. Lomolino. 1998. *Biogeography*, 2nd ed. Sunderland, MA: Sinauer.

Bruner, A. G., R. E. Gullison, R. E. Rice, and G. A. B. da Fonseca. 2001. Effectiveness of parks in protecting tropical biodiversity. *Science* 291:125–128.

Buechner, M. 1987. Conservation in insular parks: Simulation models of factors affecting the movement of animals across park boundaries. *Biological Conservation* 41:57–76.

Burbidge, A. A., and B. F. J. Manly. 2002. Mammal extinctions on Australian islands: Causes and conservation implications. *Journal of Biogeography* 29:465–473.

Burgess, E. W. 1924. The growth of the city: An introduction to a research project. *American Sociological Society Publications* 18:85–97.

Burke, D. M., and E. Nol. 1998. Influence of food abundance, nest-site habitat, and forest fragmentation on breeding ovenbirds. *Auk* 115:96–104.

Burkey, T. V. 1997. Metapopulation extinction in fragmented landscapes: Using bacteria and protozoa communities as model ecosystems. *American Naturalist* 150:568–591.

Burnett, J. 1998. Controversial bio-corridor seen as way to preserve Central American environment. *Sustainable development reporting project.* National Public Radio. http://lanic.utexas.edu/project/sdrp/corridor.html.

Buza, L., A. Young, and P. Thrall. 2000. Genetic erosion, inbreeding and reduced fitness in fragmented populations of the endangered tetraploid pea *Swainsona recta. Biological Conservation* 93:177–186.

Cabeza, M., and A. Moilanen. 2001. Design of reserve networks and the persistence of biodiversity. *Trends in Ecology and Evolution* 16:242–248.

Calabrese, J. M., and W. F. Fagan. 2004. A comparison-shopper's guide to connectivity metrics. *Frontiers in Ecology and the Environment* 2:529–536.

Canadian Press. 2005. Bear paths protect neither man nor beast. June 12.

Canadian Parks and Wilderness Society. 2005. The Cascade corridor: An experiment in restoring connectivity. In *What's happening in our mountain parks?* CPAWS Calgary/Banff Chapter, Calgary. http://www.cpawscalgary.org/mountain-parks/cascade-corridor.html.

Cantrell, R. S., C. Cosner, and W. F. Fagan. 1998. Competition reversals inside ecological reserves: The role of external habitat degradation. *Journal of Mathematical Biology* 37:491–533.

Carey, A. B., and M. L. Johnson. 1995. Small mammals in managed, naturally young, and old-growth forests. *Ecological Applications* 5:336–352.

Carignan, V., and M. A. Villard. 2002. Selecting indicator species to monitor ecological integrity: A review. *Environmental Monitoring and Assessment.* 78:45–61.

Caro, T. M., and G. O'Doherty. 1999. On the use of surrogate species in conservation biology. *Conservation Biology* 13:805–814.

Carroll, C., M. K. Phillips, N. H. Schumaker, and D. W. Smith. 2003. Impacts of landscape change on wolf restoration success: Planning a reintroduction program based on static and dynamic spatial models. *Conservation Biology* 17:536–548.

Carvalho, K. S., and H. L. Vasconcelos. 1999. Forest fragmentation in central Amazonia and its effects on litter-dwelling ants. *Biological Conservation* 91:151–157.

Castro, R. L., I. Gamez, N. Olson, and F. Tattenbach. 1998. *The Costa Rican experience with market instruments to mitigate climate change and conserve biodiversity.* San Jose, Costa Rica: Fundacion para el Desarollo de la Cordillera Volcanica Central, Ministerio de Ambiente y Energia, and World Bank.

Cat Specialist Group. 2000. *Panthera leo* ssp. *persica.* 2004 IUCN Red List of Threatened Species. http://www.redlist.org/search/details.php?species=15952.

Cestero, B. 1999. *Beyond the hundredth meeting: A field guide to collaborative conservation on the West's public lands.* Tucson: Sonoran Institute.

Chalfoun, A. D., F. R. Thompson III, and M. J. Ratnaswamy. 2002. Nest predators and fragmentation: A review and meta-analysis. *Conservation Biology* 16:306–318.

Chapin, F. S., III, B. H. Walker, R. J. Hobbs, D. U. Hooper, J. H. Lawton, O. E. Sala, and D. Tilman. 1997. Biotic control over the functioning of ecosystems. *Science* 277:500–504.

Chomitz, K., and D. Gray. 1996. Roads, land use and deforestation: A spatial model applied to Belize. *World Bank Economic Review* 10:487–512.

Church, R., D. Stoms, and F. Davis. 1996. Reserve selection as a maximal coverage location problem. *Biological Conservation* 76:105–112.

Churcher, J. B., and J. H. Lawton. 1987. Predation by domestic cats in an English village. *Journal of Zoology* 212:439–456.

Clements, F. E. 1897. Peculiar zonal formations of the Great Plains. *American Naturalist* 31:968–970.

Clevenger, A. P., B. Chruszez, and K. E. Gunson. 2001. Highway mitigation fencing reduces wildlife-vehicle collisions. *Wildlife Society Bulletin* 29:646–653.

Clevenger, A. P., and N. Waltho. 2005. Performance indices to identify attributes of highway crossing structures facilitating movement of large mammals. *Biological Conservation* 121:453–464.

Cockburn, A. 2003. Cooperative breeding in oscine passerines: Does sociality inhibit speciation? *Proceedings of the Royal Society of London, Series B: Biological Sciences* 270:2207–2214.

Cohen, A. N., and J. T. Carlson. 1998. Accelerating invasion rate in a highly invaded estuary. *Science* 279:555–558.

Collinge, S. K. 2000. Effects of grassland fragmentation on insect species loss, colonization, and movement patterns. *Ecology* 81:2211–2226.

Compton, J. E., and R. D. Boone. 2000. Long-term impacts of agriculture on soil carbon and nitrogen in New England forests. *Ecology* 81:2314–2330.

Conservation Biology Institute. 2002. *Wildlife corridor monitoring study: Multiple species conservation plan.* Sacramento: California Department of Fish and Game.

Cook, E. A., and H. N. van Lier, eds. 1994. *Landscape planning and ecological networks.* Amsterdam: Elsevier Science.

Cook, J. E. 1996. Implications of modern successional theory for habitat typing: A review. *Forest Science* 42:67–75.

Coppolillo, P., H. Gomez, F. Maisels, and R. Wallace. 2004. Selection criteria for suites of landscape species as a basis for site-based conservation. *Biological Conservation* 115:419–430.

Costanza, R., R. d'Arge, R. de Groot, S. Farber, M. Grasso, B. Hannon, K. Limburg et al. 1997. The value of the world's ecosystem services and natural capital. *Nature* 387:253–260.

Costello, C., and S. Polasky. 2004. Dynamic reserve site selection. *Resource and Energy Economics* 26:157–174.

Cova, T. J., and J. P. Johnson. 2002. Microsimulation of neighborhood evacuations in the urban-wildland interface. *Environment and Planning A* 34:2211–2229.

Cox, P. A., and T. Elmquist. 2000. Pollinator extinction in the Pacific Islands. *Conservation Biology* 14:1237–1239.

Cox, P. A., T. Elmquist, E. D. Pierson, and E. D. Rainey. 1991. Flying foxes as strong interactors in South Pacific Island ecosystems: A conservation hypothesis. *Conservation Biology* 5:448–454.

Craighead, J. J. 1991. Yellowstone in transition. In *The Greater Yellowstone Ecosystem: Redefining America's wilderness heritage*, eds. R. B. Kieter and M. S. Boyce, pages 27–40. New Haven, CT: Yale University Press.

Cramer, P. C., and K. M. Portier. 2001. Modeling Florida panther movements in response to human attributes of the landscape and ecological settings. *Ecological Modelling* 140:51–80.

Crome, F., J. Isaacs, and L. Moore. 1994. The utility to birds and mammals of remnant riparian vegetation and associated windbreaks in the tropical Queensland uplands. *Pacific Conservation Biology* 1:328–343.

Crooks, K. R., and M. E. Soulé. 1999. Mesopredator release and avifaunal extinctions in a fragmented system. *Nature* 400:563–566.

Crooks, K. R., A. V. Suarez, D. T. Bolger, and M. E. Soulé. 2001. Extinction and colonization of birds on habitat islands. *Conservation Biology* 15:159–172.

Cross, H. C., P. D. Wettin, and F. M. Keenan. 1991. Corridors for wetland conservation and management? Room for conjecture. In *Nature conservation 2: The role of corridors*, eds. D. A. Saunders and R. J. Hobbs, pages 159–165. Chipping Norton, New South Wales, Australia: Surrey Beatty & Sons.

Csuti, B., S. Polasky, P. H. Williams, R. L. Pressey, J. D. Camm, M. Kershaw, A. R. Kiester et al. 1997. A comparison of reserve selection algorithms using data on terrestrial vertebrates in Oregon. *Biological Conservation* 80:83–97.

Daily, G. C., ed. 1997. *Nature's services: Societal dependence on natural ecosystems*. Washington, DC: Island Press.

Daily, G. C., S. Alexander, P. R. Ehrlich, L. Goulder, J. Lubchenco, P. A. Matson, H. A. Mooney et al. 1997. Ecosystem services: Benefits supplied to human societies by natural ecosystems. *Issues in Ecology*, no. 2:1–16.

Daily, G. C., and K. Ellison. 2002. *The new economy of nature: The quest to make conservation profitable*. Washington, DC: Island Press.

Dalby, A. 2003. *Language in danger: The loss of linguistic diversity and the threat to our future*. New York: Columbia University Press.

Dale, S. 2001. Female-biased dispersal, low female recruitment, unpaired males, and the extinction of small and isolated bird populations. *Oikos* 92:344–356.

Dale, S., K. Mork, R. Solvang, and A. J. Plumptre. 2000. Edge effects on the understory bird community in a logged forest in Uganda. *Conservation Biology* 14:265–276.

Danielson, B. J. 1991. Communities in a landscape: The influence of habitat heterogeneity on the interactions between species. *American Naturalist* 138:1105–1120.

Danielson, B. J., and M. W. Hubbard. 2000. The influence of corridors on the movement behavior of individual *Peromyscus polionotus* in experimental landscapes. *Landscape Ecology* 15:323–331.

D'Antonio, C. M., T. L. Dudley, and M. Mack. 1999. Disturbance and biological invasions: Direct effects and feedbacks. In *Ecosystems of disturbed ground*, vol. 16 of *Ecosystems of the world*, ed. L. R. Walker, pages 413–452. Amsterdam: Elsevier Science.

Darling, F. F. 1938. *Bird flocks and the breeding cycle: A contribution to the study of avian sociality*. Cambridge: Cambridge University Press.

Date, E. M., H. A. Ford, and H. F. Recher. 1991. Frugivorous pigeons, stepping stones, and weeds in northern New South Wales. In *Nature conservation 2: The role of corridors*, eds. D. A. Saunders and R. J. Hobbs, pages 141–145. Chipping Norton, New South Wales, Australia: Surrey Beatty & Sons.

Dawson, K. J. 1995. A comprehensive conservation strategy for Georgia's greenways. *Landscape and Urban Planning* 33:27–43.

DeChaine, E. G., and A. P. Martin. 2004. Historic cycles of fragmentation and expansion in *Parnassius smintheus* (Papilionidae) inferred using mitochondrial DNA. *Evolution* 58:113–127.

Delattre, P., B. De Sousa, E. Fichet-Calvet, J. P. Quéré, and P. Giraudoux. 1999. Vole outbreaks in a landscape context: Evidence from a six-year study of *Microtus arvalis*. *Landscape Ecology* 14:401–412.

Delattre, P., P. Giraudoux, J. Baudry, P. Musard, M. Toussaint, D. Truchetet, P. Stahl et al. 1992. Land use patterns and types of common vole (*Microtus arvalis*) population kinetics. *Agriculture, Ecosystems and Environment* 39:153–168.

Delattre, P., P. Giraudoux, J. Baudry, J. P. Quéré, and E. Fichet. 1996. Effect of landscape structure on common vole (*Microtus arvalis*) distribution and abundance at several spatial scales. *Landscape Ecology* 11:279–288.

Denstan, C. E., and B. J. Fox 1996. The effects of fragmentation and disturbance of rainforest on ground-dwelling small mammals on the Robertson Plateau, New South Wales, Australia. *Journal of Biogeography* 23:187–201.

Diamond, J. M. 1972. Biogeographic kinetics: Estimation of relaxation times for avifaunas of southwest Pacific Islands. *Proceedings of the National Academy of Sciences* 69:3199–3203.

Diamond, J. M. 1973. Distributional ecology of New Guinea birds. *Science* 179:759–769.

Diamond, J. M. 1976. Island biogeography and conservation: Strategies and limitations. *Science* 193:1027–1029.

Dijak, W. D., and F. R. Thompson. 2000. Landscape and edge effects on the distribution of mammalian predators in Missouri. *Journal of Wildlife Management* 64:209–216.

Dobson, A. 2003. Metalife! *Science* 301:1488–1490.

Dobson, A. P., and A. Merenlender. 1991. Coevolution of macroparasites and their hosts. In *Parasite-host associations: Coexistence or conflict?* eds. C. A. Toft, A. Aeschlimann, and L. Bolis, pages 83–101. New York: Oxford University Press.

Donald, P. F., R. E. Green, and M. F. Heath. 2001. Agricultural intensification and the collapse of Europe's farmland bird populations. *Proceedings of the Royal Society of London, Series B: Biological Sciences* 268:25–29.

Donohue K., D. R. Foster, and G. Motzkin. 2000. Effects of the past and the present on species distribution: Land-use history and demography of wintergreen. *Journal of Ecology* 88:303–316.

Donovan, T. M., and C. H. Flather. 2002. Relationships among North American songbird trends, habitat fragmentation, and landscape connectivity. *Ecological Applications* 12:364–374.

Donovan, T. M., and R. H. Lamberson. 2001. Area-sensitive distributions counteract negative effects of habitat fragmentation on breeding birds. *Ecology* 82:1170–1179.

Downes, S. J., K. A. Handasyde, and M. A. Elgar. 1997. The use of corridors by mammals in fragmented Australian eucalypt forests. *Conservation Biology* 11:718–726.

Drapeau, P., A. Leduc, J. Giroux, J. L. Savard, Y. Bergeron, and W. L. Vickery. 2000. Landscape-scale disturbances and changes in bird communities of boreal mixed-wood forests. *Ecological Monographs* 70:423–444.

Driscoll, M. J. L., and T. M. Donovan. 2004. Landscape context moderates edge effects: Nesting success of wood thrushes in central New York. *Conservation Biology* 18:1330–1338.

Duane, T. P. 1997. Community participation in ecosystem management. *Ecology Law Quarterly* 24.

Dubost, F. 1998. De la maison de campagne à la résidence secondaire. In *L'autre maison: La "résidence secondaire," refuge des générations*, ed. F. Dubost, pages 10–37. Paris: Autrement.

Duelli, P., M. Studer, I. Marchand, and S. Jakob. 1990. Population movements of arthropods between natural and cultivated areas. *Biological Conservation* 54:193–207.

Duke, D. L., M. Hebblewhite, P. C. Paquet, C. Callaghan, and M. Percy. 2001. Restoring a large-carnivore corridor in Banff National Park. In *Large mammal restoration: Ecological and sociological challenges in the 21st century*, eds. D. S. Maehr, R. F. Noss, and J. L. Larkins, pages 261–275. Washington, DC: Island Press.

Dunn, R. R. 2000. Isolated trees as foci of diversity in active and fallow fields. *Biological Conservation* 95:317–321.

Dunning, J. B., Jr., R. Borgella Jr., K. Clements, and G. K. Meffe. 1995. Patch isolation, corridor effects, and colonization by a resident sparrow in a managed pine woodland. *Conservation Biology* 9:542–550.

Dunning, J. B., B. J. Danielson, and H. R. Pulliam. 1992. Ecological processes that affect populations in complex landscapes. *Oikos* 65:169–175.

Dybas, C. L. 2007. California wine country clashes with ecosystem. *Washington Post*, June 21, page A08.

Edmands, S. 1999. Heterosis and outbreeding depression in interpopulation crosses spanning a wide range of divergence. *Evolution* 53:1757–1768.

Edwards, T. C., E. T. Deshler, D. Foster, and G. G. Moisen. 1996. Adequacy of wildlife habitat relation models for estimating spatial distributions of terrestrial vertebrates. *Conservation Biology* 10:263–270.

Ehrlich, P., and A. Ehrlich. 2004. *One with Nineveh: Politics, consumption, and the human future*. Washington, DC: Island Press.

Ellstrand, N. C., and K. A. Schierenbeck. 2000. Hybridization as a stimulus for the evolution of invasiveness in plants? *Proceedings of the National Academy of Sciences* 97:7043–7050.

Environmental Protection Agency. 2002. *Great Plains program: A phase I inventory of the current EPA efforts to protect ecosystems*. Washington, DC: Environmental Protection Agency, Office of Wetlands, Oceans, and Watersheds.

Epps, C. W., D. R. McCullough, J. D. Wehausen, V. C. Bleich, and J. L. Rechel. 2004. Effects of climate change on population persistence of desert-dwelling mountain sheep in California. *Conservation Biology* 18:102–113.

Éri, V. 2001. *Sprawl or smart urban growth?* [in Hungarian]. Budapest: Center for Environmental Studies.

Estes, J. A., and D. O. Duggins. 1995. Sea otters and kelp forest in Alaska: Generality and variation in a community ecological paradigm. *Ecological Monographs* 65:75–100.

European Center for Nature Conservation and Council of Europe. 2003. *Pan-European Ecological Network fact sheet 7: Coastal and marine networks*. Kiev, Ukraine: Authors.

Evans, F. C. 1942. Studies of a small mammal population in Bagley Wood, Berkshire. *Journal of Animal Ecology* 11:194–197.

Fábos, J. G. 2004. Greenway planning in the United States: Its origins and recent case studies. *Landscape and Urban Planning* 68:321–342.

Fahrig, L. 2003. Effect of habitat fragmentation on biodiversity. *Annual Review of Ecology, Evolution and Systematics* 34:487–515.

Fahrig, L., and G. Merriam. 1994. Conservation of fragmented populations. *Conservation Biology* 8:50–59.

Fairbanks, W. S., and R. Tullous. 2002. Distribution of pronghorn (*Antilocapra americana* Ord) on Antelope Island State Park, Utah, USA, before and after establishment of recreational trails. *Natural Areas Journal* 22:277–282.

Fairfax, S. K. and D. Guenzler. 2001. *Conservation trusts*. Lawrence: University Press of Kansas.

Fairfax, S. K., L. Reymond, L. Gwin, M. A. King, and L. A. Watt. 2005. *Buying Nature: The Limits of land acquisition as a conservation strategy.* Boston: MIT Press.

Faith, D. P., and P. A. Walker. 2002. The role of trade-offs in biodiversity conservation planning: Linking local management, regional planning and global conservation efforts. *Journal of Biosciences* 27:393–407.

Farris, K. L., M. J. Huss, and S. Zack. 2004. The role of foraging woodpeckers in the decomposition of ponderosa pine snags. *The Condor* 106:50–59.

Fausch, K. D., J. Lyons, J. R. Karr, and P. L. Angermeier. 1990. Fish communities as indicators of environmental degradation. In *American Fisheries Symposium*, ed. S. M. Adams, pages 123–145. Bethesda, MD: American Fisheries Society.

Fearnside, P. M. 1993. Deforestation in the Brazilian Amazon: The effect of population and land tenure. *Ambio* 8:537–545.

Fedriani, J. M., T. K. Fuller, R. M. Sauvajot, and E. C. York. 2000. Competition and intraguild predation among three sympatric carnivores. *Oecologia* 125:258–270.

Fernández-Juricic, E. 2001. Density-dependent habitat selection of corridors in a fragmented landscape. *Ibis* 143:278–287.

Ferraro, P. J. 2001. Global habitat protection: Imitations of development interventions and a role for conservation performance payments. *Conservation Biology* 15:990–1000.

Ferraro, P. J., and A. Kiss. 2002. Ecology: Direct payments to conserve biodiversity. *Science* 298:1718–1719.

Ferraz, G., G. J. Russell, P. C. Stouffer, R. O. Bierregaard Jr., S. L. Pimm, and T. E. Lovejoy. 2003. Rate of species loss from Amazonian forest fragments. *Proceedings of the National Academy of Sciences* 100:14069–14073.

Ferren, W. R. J., and D. M. Hubbard. 1996. Review of ten years of vernal pool restoration and creation in Santa Barbara, California. In *Ecology, conservation, and management of vernal pool ecosystems*, eds. C. W. Witham, E. T. Bauder, D. Belk, F. W. R. Ferren Jr., and R. Ornduff, pages 206–216. Sacramento: California Native Plant Society.

Ferrier, S. 2002. Mapping spatial pattern in biodiversity for regional conservation planning: Where to from here? *Systematic Biology* 51:331–363.

Ferson, S., and H. R. Akçakaya. 1990. *RAMAS/age user manual*. Setauket, NY: Applied Biomathematics.

Fleishman, E., D. D. Murphy, and R. B. Blair. 2001. Selecting effective umbrella species. *Conservation Biology in Practice* 2:17–23.

Fleury, A. M., and R. D. Brown. 1997. A framework for the design of wildlife conservation corridors with specific application to southwestern Ontario. *Landscape and Urban Planning* 37:163–186.

Floursheim, J. L., and J. F. Mount. 2002. Restoration of floodplain topography by sandsplay complex formation in response to intentional levee breaches, Lower Consumnes River, California. *Geomorphology* 44:67–94.

Folke, C., S. Carpenter, B. Walker, M. Scheffer, T. Elmqvist, L. H. Gunderson, and C. S. Holling. 2004. Regime shifts, resilience, and biodiversity in ecosystem management. *Annual Review of Ecology and Systematics* 35:557–581.

Fontaine, C., and A. Gonzalez. 2005. Population synchrony induced by resource fluctuations and dispersal in an aquatic microcosm. *Ecology* 86:1463–1471.

Food and Agriculture Organization of the United Nations. 2003. *State of the world's forests.* Rome: Author.

Ford, H. A., G. W. Barrett, D. A. Saunders, and H. F. Recher. 2001. Why have birds in the woodlands of Southern Australia declined? *Biological Conservation* 97:71–88.

Foreman, D. 1999. The wildlands project and the rewilding of North America. *Denver University Law Review* 76:535–553.

Forman, R. T. T. 1991. Landscape corridors: From theoretical foundations to public policy. In *Nature conservation 2: The role of corridors,* eds. D. A. Saunders and R. J. Hobbs, pages 71–84. Chipping Norton, New South Wales, Australia: Surrey Beatty & Sons.

Forman, R. T. T. 1995. *Land mosaics: The ecology of landscapes and regions.* Cambridge: University of Cambridge.

Forman, R. T. T., and M. Godron. 1986. *Landscape ecology.* New York: John Wiley.

Forman, R. T. T., D. Sperling, J. A. Bissonette, A. P. Clevenger, C. D. Cutshall, V. H. Dale, L. Fahrig et al. 2003. *Road ecology.* Washington, DC: Island Press.

Forys, E. A., and S. R. Humphrey. 1996. Home range and movements of the lower keys marsh rabbits in a highly fragmented habitat. *Journal of Mammalogy* 77:1042–1048.

Foster, D. R., G. Motzkin, and S. Slater. 1998. Land-use history as long-term broad-scale disturbance: Regional forest dynamics in central New England. *Ecosystems* 1:96–119.

Foster, M. L., and S. R. Humphrey. 1995. Use of highway underpasses by Florida panthers and other wildlife. *Wildlife Society Bulletin* 23:95–100.

Friedman, D. S. 1997. Walking on the wild side. *Landscape Architecture,* September, 52–57.

Fritts, T. H., and G. H. Rodda. 1998. The role of introduced species in the degradation of island ecosystems: A case history of Guam. *Annual Review of Ecology and Systematics* 29:113–140.

Fritz, R., and G. Merriam. 1993. Fencerow habitats for plants moving between farmland forests. *Biological Conservation* 64:141–148.

Fry, G. L. A. 1994. The role of field margins in the landscape. *BCPC Monograph. Field Margins: Integrating Agriculture and Conservation* 58:31–39.

Galle, L., K. Margoczi, E. Kovacs, G. Gyorffy, L. Kormoczi, and L. Neneth. 1995. River valleys: Are they ecological corridors? *Tiscia* [Szeged, Hungary] 29:53–58.

Gardner, R. H., B. T. Milne, M. G. Turner, and R. V. O'Neil. 1987. Neutral models for the analysis of broad-scale landscape pattern. *Landscape Ecology* 1:19–28.

Garrison, B. A., R. A. Erickson, M. A. Patten, and I. C. Timossi. 2000. Accuracy of wildlife model predictions for bird species occurrences in California counties. *Wildlife Society Bulletin* 28:667–674.

Garrod, G., and K. Willis. 1999. *Economic valuation of the environment: Methods and case studies.* Cheltenham, England: Edward Elgar.

Gascon, C., T. E. Lovejoy, R. O. Bierregaard Jr., J. R. Malcolm, P. C. Stouffer, H. L. Vasconcelos, W. F. Laurance, B. Zimmerman, M. Tocher, and S. Borges. 1999. Matrix habitat and species richness in tropical forest remnants. *Biological Conservation* 91:223–229.

Gehlbach, F. R., and R. S. Baldridge. 1987. Live blind snakes (*Leptotyphlops dulcis*) in eastern screech owl (*Otus asio*) nests: A novel commensalism. *Oecologia* 71:560–563.

Gehring, T. M., and R. K. Swihart. 2003. Body size, niche breadth, and ecologically scaled responses to habitat fragmentation: Mammalian predators in an agricultural landscape. *Biological Conservation* 109:283–295.

Gharrett, A. J., and W. W. Smoker. 1991. Two generations of hybrids between even- and odd-year pink salmon (*Onchorhynchus gorbuscha*): A test for outbreeding depression. *Canadian Journal of Fish and Aquatic Science* 48:1744–1749.

Gibeau, M. L., A. P. Clevenger, S. Herrero, and J. Wierzchowski. 2002. Grizzly bear response to human development and activities in the Bow River Watershed, Alberta, Canada. *Biological Conservation* 103:227–236.

Gilbert, F., A. Gonzalez, and I. Evans-Freke. 1998. Corridors maintain species richness in the fragmented landscapes of a microecosystem. *Proceedings of the Royal Society of London, Series B: Biological Sciences* 265:577–582.

Gilpin, M. 1991. The genetic effective size of a metapopulation. *Biological Journal of the Linnean Society* 42:165–175.

Gilpin, M. E., and M. E. Soulé. 1986. Minimum viable populations: Processes of species extinction. In *Conservation biology: The science of scarcity and diversity*, ed. M. E. Soulé, pages 19–34. Sunderland, MA: Sinauer.

Giraudoux, P., D. A. Vuitton, S. Bresson-Hadni, P. Craig, B. Bartholomot, G. Barnish, J. J. Laplante, S. D. Zhong, and D. Lenys. 1996. Mass screening and epidemiology of alveolar echinococcosis in France, western Europe, and Gansu, central China: From epidemiology towards transmission ecology. In *Alveolar echinococcosis: Strategy for eradication of alveolar echinococcosis of the liver*, eds. J. Uchino and N. Sato, pages 197–211. Sapporo, Japan: Fuji Shoin.

Giusti, G. A., and A. M. Merenlender. 2002. Inconsistent application of environmental laws and policies to California's oak woodlands. In *Fifth Symposium on Oak Woodlands*, Gen. Tech. Rep., eds. R. Standiford, D. McCreary, and K. L. Purcell, pages 473–482. San Diego, CA: USDA Forest Service.

Gog, J., R. Woodroffe, and J. Swinton. 2002. Disease in endangered metapopulations: The importance of alternative hosts. *Proceedings of the Royal Society of London, Series B: Biological Sciences* 269:671–676.

Goldman, M. 2003. Partitioned nature, privileged knowledge: Community-based conservation in Tanzania. *Development and Change* 34:833–862.

Gonzalez, A. 2000. Community relaxation in fragmented landscapes: The relation between species richness, area and age. *Ecology Letters* 3:441–448.

Gosz, J. R. 1993. Ecotone hierarchies. *Ecological Applications* 3:369–376.

Gosz, J. R., and P. J. H. Sharpe. 1989. Broad-scale concepts for interactions of climate, topography, and biota at biome transitions. *Landscape Ecology* 3:229–243.

Grenfell, B. T., B. M. Bolker, and A. Kleczkowski. 1995. Seasonality and extinction in chaotic metapopulations. *Proceedings of the Royal Society of London, Series B: Biological Sciences* 259:97–103.

Grenfell, B., and J. Harwood. 1997. (Meta)population dynamics of infectious diseases. *Trends in Ecology and Evolution* 12:395–399.

Griffith, B., J. M. Scott, J. W. Carpenter, and C. Reed. 1989. Translocation as a species conservation tool: Status and strategy. *Science* 245:477–480.

Grimm, V. 1999. Ten years of individual-based modelling in ecology: What have we learned and what could we learn in the future? *Ecological Modelling* 115:129–148.

Groom, M. J. 1998. Allee effect limits population viability of an annual plant. *American Naturalist* 151:487–496.

Groves, C. R. 2003. *Drafting and conservation blueprint: A practitioner's guide to planning for biodiversity.* Washington, DC: Island Press.

Groves, C. R., D. B. Jensen, L. L. Valutis, K. H. Redford, M. L. Shaffer, J. M. Scott, J. V. Baumgartner, J. V. Higgins, M. W. Beck, and M. G. Anderson. 2002. Planning for biodiversity conservation: Putting conservation science into practice. *BioScience* 52:499–512.

Gu, W. D., and R. K. Swihart. 2004. Absent or undetected? Effects of non-detection of species occurrence on wildlife-habitat models. *Biological Conservation* 116:195–203.

Guichard, F., S. A. Levin, A. Hastings, and D. Siegel. 2004. Toward a dynamic metacommunity approach to marine reserve theory. *BioScience* 54:1003–1011.

Guisan, A., and N. E. Zimmermann. 2000. Predictive habitat distribution models in ecology. 135:147–186.

Gunderson, L. H., L. J. Pritchard, C. S. Holling, C. Folke, and G. D. Peterson. 2002. A summary and synthesis of resilience in large-scale systems. In *Resilience and the behavior of large-scale systems*, eds. L. H. Gunderson and L. J. Prichard, page 287. Washington, DC: Island Press.

Gustanski, J. A., and R. H. Squires. 2000. *Protecting the land: Conservation easements past, present, and future.* Washington, DC: Island Press.

Gyorgy, T. 1994. The role of remote sensing in the protection of the environment and nature. *Erdeszeti es Faipari Tudomanyos Kozlemenyek* (Hungary) 2:139–146.

H&N&S landschapsarchitecten. 1996. *Over scherven en gelule: Een rapport over de versnippering nan de natuur in Nederland.* Rapport LNV/VenW/VRom, The Hague.

Haas, C. A. 1995. Dispersal and use of corridors by birds in wooded patches on an agricultural landscape. *Conservation Biology* 9:845–854.

Haddad, N. M. 1999. Corridor use predicted from behaviors at habitat boundaries. *American Naturalist* 153:215–227.

Haddad, N. M., and J. J. Tewksbury. 2005. Low-quality habitat corridors as movement conduits for two butterfly species. *Ecological Applications* 15:250–257.

Haight, R. G., S. A. Snyder, and C. S. Revelle. 2005. Metropolitan open-space protection with uncertain site availability. *Conservation Biology* 19:327–337.

Hailey, T. L., and R. DeArment. 1972. Droughts and fences restrict pronghorns. *Proceedings of the Biennial Pronghorn Antelope Workshop* 5:22–24.

Haines, J. 2006. FWP moves bears out of town. *Bozeman* (Montana) *Daily Chronicle*, October 8, 2004.

Hall, J. S., D. J. Harris, V. Medjibe, and P. M. S. Ashton. 2003. The effects of selective logging on forest structure and tree species composition in a Central African forest: Implications for management of conservation areas. *Forest Ecology and Management* 183:249–264.

Hamer, T. E., E. D. Forsman, A. D. Fuchs, and M. L. Walters. 1994. Hybridization between barred and spotted owls. *Auk* 111:487–492.

Hannon, S. J., C. A. Paszkowski, S. J. DeGroot, S. E. Macdonald, M. Wheatley, and B. R. Eaton. 2002. Abundance and species composition of amphibians, small mammals, and songbirds in riparian forest buffer strips of varying widths in the boreal forest of Alberta. *Canadian Journal of Forest Research* 32:1784–1800.

Hannon, S. J., and F. K. A. Schmiegelow. 2002. Corridors may not improve the conservation value of small reserves for most boreal birds. *Ecological Applications* 12:1457–1468.

Hansen, A. J., R. Rasker, B. Maxwell, J. J. Rotella, J. D. Johnson, A. Wright Parmenter, U. Langner, W. B. Cohen, R. L. Lawrence, and M. P. V. Kraska. 2002. Ecological causes and consequences of demographic change in the new west. *BioScience* 52:151–162.

Hansen, A. J., and J. J. Rotella. 2002. Biophysical factors, land use, and species viability in and around nature reserves. *Conservation Biology* 16:1112–1122.

Hanski, I. 1999. *Metapopulation ecology*. Oxford: Oxford University Press.

Hanski, I., J. Alho, and A. Moilanen. 2000. Estimating the parameters of survival and migration of individuals in metapopulations. *Ecology* 8:239–251.

Hanski, I., and M. E. Gilpin. 1997. *Metapopulation biology: Ecology, genetics and evolution*. San Diego, CA: Academic Press.

Hanski, I., L. Hansson, and H. Henttonen. 1991. Specialist predators, generalist predators, and the microtine rodent cycle. *Journal of Animal Ecology* 60:353–367.

Hanson, J. S., G. P. Malanson, and M. P. Armstrong. 1990. Landscape fragmentation and dispersal in a model of riparian forest dynamics. *Ecological Modelling* 49:277–296.

Hansson, L. 1995. Development and application of landscape approaches in mammalian ecology. In *Landscape approaches in mammalian ecology and conservation*, ed. W. Z. Lidicker Jr., pages 20–39. Minneapolis: University of Minnesota Press.

Hansson, L., L. Fahrig, and G. Merriam, eds. 1995. *Mosaic landscapes and ecological processes*. London: Chapman and Hall.

Haroldson, M. A., and K. Frey. 2002. Grizzly bear mortalities. In *Yellowstone grizzly bear investigations: Annual report of the Interagency Study Team, 2001*, eds. C. C. Schwartz and M. A. Haroldson, pages 23–28. Bozeman, MT: United States Geological Survey.

Harris, L. D., T. Hoctor, D. Maehr, and J. Sanderson. 1996. The role of networks and corridors in enhancing the value and protection of parks and equivalent areas. In *National parks and protected areas*, ed. R. G. Wright, pages 173–197. Cambridge, England: Blackwell Science.

Harris, L. D., and J. Scheck. 1991. From implications to applications: The dispersal corridor principle applied to the conservation of biological diversity. In *Nature conservation 2: The role of corridors*, eds. D. A. Saunders and R. J. Hobbs, pages 189–220. Chipping Norton, New South Wales, Australia: Surrey Beatty & Sons.

Harrison, S. 1991. Local extinction in a metapopulation context: An empirical evaluation. *Biological Journal of the Linnean Society* 42:73–88.

Harrison, S. 1999. Local and regional diversity in a patchy landscape: Native, alien, and endemic herbs on serpentine. *Ecology* 80:70–80.

Harte, J. 2001. Land use, biodiversity, and ecosystem integrity: The challenge of preserving Earth's life support system. *Ecology Law Quarterly* 27:929–965.

Hartley, M. J., and M. L. Hunter Jr. 1998. A meta-analysis of forest cover, edge effects, and artifical nest predation rates. *Conservation Biology* 12:465–469.

Hastings, A. 2003. Metapopulation persistence with age-dependent disturbance or succession. *Science* 301:1525–1526.

Havlick, D. 2004. Roadkill. *Conservation in Practice*, Winter, 30–34.

Hawkins, B. A., M. B. Thomas, and M. E. Hockberg. 1993. Refuge theory and biological control. *Science* 262:1429–1432.

Hawkins, C. C., W. E. Grant, and M. T. Longnecker. 1999. Effects of subsidized cats on California birds and rodents. *Transactions of the Western Section of the Wildlife Society* 35:29–33.

Hedrick, P. W. 1996. Genetics of metapopulations: Aspects of a comprehensive perspective. In *Metapopulations and wildlife conservation*, ed. D. R. McCullough, pages 29–51. Washington, DC: Island Press.

Henein, K., and G. Merriam. 1990. The elements of connectivity where corridor quality is variable. *Landscape Ecology* 4:157–170.

Henle, K., D. B. Lindenmayer, C. R. Margules, D. A. Saunders, and C. Wissel. 2004. Species survival in fragmented landscapes: Where are we now? *Biodiversity and Conservation* 13:1–8.

Henry, A. C., Jr., D. A. Hosack, C. W. Johnson, R. Rol, and G. Bentrup. 1999. Conservation corridors in the United States: Benefits and planning guidelines. *Journal of Soil and Water Conservation* 54:645–650.

Herre, E. A. 1993. Population structure and the evolution of virulence in nematode parasites of fig wasps. *Science* 259:1442–1445.

Heske, E. J. 1995. Mammalian abundances on forest-farm edges versus forest interiors in southern Illinois: Is there an edge effect? *Journal of Mammalogy* 76:562–568.

Hess, G. R. 1994. Conservation corridors and contagious disease: A cautionary note. *Conservation Biology* 8:256–262.

Hess, G. 1996. Disease in metapopulation models: Implications for conservation. *Ecology* 77:1617–1632.

Hess, G. R., and R. A. Fischer. 2001. Communicating clearly about conservation corridors. *Landscape and Urban Planning* 55:195–208.

Hilson, G. 2002. The environmental impact of small-scale gold mining in Ghana: Identifying problems and possible solutions. *Geographical Journal* 168:57–72.

Hilty, J. A. 2001. Use of riparian corridors by wildlife in the oak woodland vineyard landscape. PhD thesis, University of California, Berkeley.

Hilty, J. A., C. Brooks, E. Heaton, and A. M. Merenlender. 2006. Forecasting the effect of land-use change on native and non-native mammalian predator distributions. *Biodiversity and Conservation*. In press.

Hilty, J. A., and A. M. Merenlender. 2000a. A comparison of covered track-plates and remotely triggered cameras. *Transactions of the Western Section of the Wildlife Society* 36:27–31.

Hilty, J. A., and A. M. Merenlender. 2000b. Faunal indicator taxa selection for monitoring ecosystem health. *Biological Conservation* 92:185–197.

Hilty, J. A., and A. M. Merenlender. 2004. Use of riparian corridors and vineyards by mammalian predators in northern California. *Conservation Biology* 18:126–135.

Hobbs, R. J. 1992. The role of corridors in conservation: Solution or bandwagon? *Trends in Ecology and Evolution* 7:389–392.

Hobbs, R. J. 2004. The Working for Water programme in South Africa: The science behind the success. *Diversity and Distributions* 10:501–503.

Hobbs, R. J., and A. J. M. Hopkins. 1991. The role of conservation corridors in a changing climate. In *Nature conservation 2: The role of corridors*, eds. D. A. Saunders and R. J. Hobbs, pages 281–290. Chipping Norton, New South Wales, Australia: Surrey Beatty & Sons.

Hobson, K. A., E. M. Bayne, and S. L. Van Wilgenburg. 2002. Large-scale conversion of forest to agriculture in the boreal plains of Saskatchewan. *Conservation Biology* 16:1530–1541.

Hodges, M. F., and D. G. Krementz. 1996. Neotropical migratory breeding bird communities in riparian forests of different widths along the Altamaha River, Georgia. *Wilson Bulletin* 108:496–506.

Hoekstra, J. M., T. M. Boucher, T. H. Ricketts, and C. Roberts. 2005. Confronting a biome crisis: Global disparities of habitat loss and protection. *Ecology Letters* 8:23–29.

Hoffmann, A. A., R. J. Hallas, J. A. Dean, and M. Schiffer. 2003. Low potential for climatic stress adaptation in a rainforest Drosophila species. *Science* 301:100–102.

Holdaway, R. N. 1999. Introduced predators and avifaunal extinctions in New Zealand. In *Extinctions in near time: Causes, contexts, and consequences*, ed. R. D. E. MacPhee, pages 189–238. New York: Plenum.

Holling, C. S., and L. H. Gunderson. 2002. Resilience and adaptive cycles. In *Panarchy: Understanding transformations in human and natural systems*, eds. L. H. Gunderson and C. S. Holling, pages 25–62. Washington, DC: Island Press.

Homewood, K., and W. Rodgers. 1991. *Maasailand ecology: Pastoralist development and wildlife conservation in Ngorongoro, Tanzania*. New York: Cambridge University Press.

House, F. 1999. *Totem salmon*. Boston: Beacon Press.

Howard, W. E., and H. E. Childs Jr. 1959. Ecology of pocket gophers with emphasis on *Thomomys bottae mewa*. *Hilgardia* 29:110–115.

Howell, J. A., and R. H. Barrett. 1998. California wildlife habitat relationships system: A test in coastal scrub and annual grassland habitats. *California Fish and Game* 84:74–87.

Huenneke, L. F., and H. A. Mooney, eds. 1989. *Grassland structure and function: California annual grassland*. Dordrecht, Netherlands: Kluwer Academic Publishers.

Huijser, M. 2004. *Evaluation of wildlife crossing structures on US Highway 93 in Montana: Task Force on Ecology and Transportation newsletter*. Spring. Washington, DC: Transportation Research Board of the National Academies.

Hulme, D., and M. Murphree. 1999. Communities, wildlife, and the "new conservation": In Africa. *Journal of International Development* 11:277–286.

Hylander, K., C. Nilsson, and T. Göthner. 2004. Effects of buffer-strip retention and clearcutting on land snails in boreal riparian forests. *Conservation Biology* 18:1052–1062.

Ingham, D. S., and M. J. Samways. 1996. Application of fragmentation and variegation models to epigaeic invertebrates in South Africa. *Conservation Biology* 10:1353–1358.

Inman, R. M., R. R. Wigglesworth, K. H. Inman, M. K. Schwartz, B. L. Brock, and J. D. Rieck. 2004. Wolverine makes extensive movements in Greater Yellowstone Ecosystem. *Northwest Science* 78:261–266.

Johnson, T. L., and J. F. Cully. 2004. Drainages as potential corridors for the spread of sylvatic plague in black-tailed prairie dogs. *American Society of Mammalogists 84th Annual Meeting Abstracts*. Arcata, CA.

Jones & Jones Architects and Landscape Architects. 2000. Wildlife crossings for US 93 from Evaro to Polson, Montana. http://www.skillings.com/web-page/wildlife.html.

Jones, J. C., M. R. Myerscough, S. Graham, and B. P. Oldroyd. 2004. Honey bee nest thermoregulation: Diversity promotes stability. *Science* 305:402–404.

Jongman, R. H. G. 1995. Nature conservation planning in Europe: Developing ecological networks. *Landscape and Urban Planning* 32:169–183.

Jongman, R. 2004. The context and concept of ecological networks. In *Ecological networks and greenways concept, design, implementation*, eds. R. Jongman and G. Pungetti, pages 7–32. Cambridge: Cambridge University Press.

Jordan, F. 2000. A reliability-theory approach to corridor design. *Ecological Modelling* 128:211–220.

Jordon, W. R., M. E. Gilpin, and J. D. Aber, eds. 1987. *Restoration ecology: A synthetic approach to ecological research.* Cambridge: Cambridge University Press.

Joyce, K. A., J. M. Holland, and C. P. Doncaster. 1999. Influences of hedgerow intersections and gaps on the movement of carabid beetles. *Bulletin of Entomological Research* 89:523–531.

Kaiser, J. 2001. Bold corridor project confronts political reality. *Science* 293:2196–2199.

Kalela, O., T. Kaponen, E. A. Lind, U. Skaren, and J. Tast. 1961. Seasonal change of habitat in the Norwegian lemming, *Lemmus lemmus* (L.). *Annals of the Academy of Science Fennicae, 4A: Biology* 55:1–72.

Kalela, O., L. Kilpeläienen, T. Koponen, and J. Tast. 1971. Seasonal differences in habitats of the Norwegian lemming, *Lemmus lemmus* (L.), in 1959 and 1960 at Kilpisjärvi, Finnish Lapland. *Annals of the Academy of Science Fennicae, 4A: Biology* 178:1–22.

Kallimanis, A. S., W. E. Kunin, J. M. Halley, and S. P. Sgardelis. 2005. Metapopulation extinction risk under spatially autocorrelated disturbance. *Conservation Biology* 19:534–546.

Karanth, K. U., and J. D. Nichols. 1998. Estimation of tiger densities in India using photographic captures and recaptures. *Ecology* 79:2852–2862.

Kareiva, P. 1987. Habitat fragmentation and the stability of predator-prey interactions. *Nature* 326:388–390.

Karr, J. R. 1981. Assessment of biotic integrity using fish communities. *Fisheries* 6:21–27.

Kautz, R. S., and J. A. Cox. 2001. Strategic habitats for biodiversity conservation in Florida. *Conservation Biology* 15:55–77.

Kay, D., J. A. Saul, and L. Telega. 2003. *Farms, communities, and collaboration: A guide to resolving farm-neighbor conflict.* 40 pages. Washington, DC: United States Department of Agriculture, Sustainable Agriculture Research and Education Program.

Keeling, M. J., and B. T. Grenfell. 1997. Disease extinction and community size: Modeling the persistence of measles. *Science* 275:65–67.

Keller, L. F., P. Arcese, J. N. M. Smith, W. M. Hochachka, and S. C. Stearns. 1994. Selection against inbred song sparrows during a natural-population bottleneck. *Nature* 372:356–357.

Kelly, J. R., and M. A. Harwell. 1990. Indicators of ecosystem recovery. *Environmental Management* 14:527–545.

Kemper, J., R. M. Cowling, and D. M. Richardson. 1999. Fragmentation of South African renosterveld shrublands: Effects on plant community structure and conservation implications. *Biological Conservation* 90:103–111.

Kennedy, D. S. 2000. Are community watershed groups effective? Confronting the thorny issue of measuring success. In *Across the great divide: Explorations in conservation and the American West*, eds. P. Brick, D. Snow, and S. Van de Wetering, pages 188–193. Washington, DC: Island Press.

Khan, J. A. 1995. Conservation and management of Gir-Lion-Sanctuary-and-National-Park, Gujarat, India. *Biological Conservation* 73:183–188.

Kilgo, J. C., D. L. Gartner, B. R. Chapman, J. B. Dunning, K. E. Franzreb, S. A. Gauthreaux, C. H. Greenberg, D. J. Levey, K. V. Miller, and S. F. Pearson. 2002. A test of an expert-based bird-habitat relationship model in South Carolina. *Wildlife Society Bulletin* 30:783–793.

Kilner, R. M., J. R. Madden, and M. E. Hauber. 2004. Brood parasitic cowbird nestlings use host young to procure resources. *Science* 305:877–879.

King, D. I., C. R. Griffin, and R. M. DeGraff. 1996. Effects of clearcutting on habitat use and reproductive success of the ovenbird in forested landscapes. *Conservation Biology* 14:1075–1087.

Kinley, T. A., and C. D. Apps. 2001. Mortality patterns in a subpopulation of endangered mountain caribou. *Wildlife Society Bulletin* 29:158–164.

Kleijn, D., F. Berendse, R. Smit, N. Gilissen, J. Smit, B. Brak, and R. Groeneveld. 2004. Ecological effectiveness of agri-environment schemes in different agricultural landscapes in the Netherlands. *Conservation Biology* 18:775–786.

Klemens, M. W. 2000. From information to action: Developing more effective strategies to conserve turtles. In *Turtle conservation*, ed. M. W. Klemens, pages 239–258. Washington, DC: Smithsonian Institution Press.

Kneitel, J. M., and T. E. Miller. 2003. Dispersal rates affect species composition in metacommunities of *Sarracenia purpurea* inquilines. *American Naturalist* 162:165–171.

Knowles, L. L. 2001. Did the Pleistocene glaciations promote divergence? Test of explicit refugia models in montane grasshoppers. *Molecular Ecology* 10:691–701.

Koenig, W. D., D. Van Vuren, and P. N. Hooge. 1996. Detectability, philopatry, and the distribution of dispersal distances in vertebrates. *Trends in Ecology and Evolution* 11:514–517.

Kozakiewicz, M., and J. Szacki. 1995. Movements of small mammals in a landscape: Patch restriction or nomadism? In *Landscape approaches in mammalian ecology and conservation*, ed. W. Z. Lidicker Jr., pages 78–94. Minneapolis: University of Minnesota Press.

Kremen C., V. Razafimahatratra, R. P. Guillery, J. Rakotomalala, A. Weiss, and J. S. Ratsisompatrarivo. 1999. Designing the Masoala National Park in Madagascar based on biological and socioeconomic data. *Conservation Biology* 13:1055–1068.

Kremen, C., and T. Ricketts. 2000. Global perspectives on pollination disruptions. *Conservation Biology* 14:1226–1228.

Kremen, C., N. M. Williams, R. L. Bugg, J. P. Fay, and R. W. Thorp. 2004. The area requirements of an ecosystem service: Crop pollination by native bee communities in California. *Ecology Letters* 7:1109–1119.

Kremen, C., N. M. Williams, and R. W. Thorp. 2002. Crop pollination from native bees at risk from agricultural intensification. *Proceedings of the National Academy of Sciences* 99:16812–16816.

Kruchek, B. L. 2004. Use of tidal marsh and upland habitats by the marsh rice rat (*Oryzomys palustris*). *Journal of Mammalogy* 85:569–575.

Kubeš J. 1996. Biocentres and corridors in a cultural landscape: A critical assessment of the territorial system of ecological stability. *Landscape and Urban Planning* 35:231–240.

Kucera, T. E., and R. H. Barrett. 1995. California wildlife faces uncertain future. *California Agriculture* 49:23–27.

Lacy, R. G., and P. S. Miller. 1993. VORTEX: A computer-simulation model for population viability analysis. *Wildlife Research* 20:45–65.

Lahaye, W. S., R. J. Gutierrez, and H. R. Akçakaya. 1994. Spotted owl metapopulation dynamics in Southern California. *Journal of Animal Ecology* 63:775–785.

Lambeck, R. J. 1997. Focal species: A multi-species umbrella for nature conservation. *Conservation Biology* 11:849–856.

Lamont, B. B., P. G. L. Klinkhamer, and E. T. F. Witkowski. 1993. Population fragmentation may reduce fertility to zero in *Banksia goodii*: A demonstration of the Allee effect. *Oecologia* 94:446–450.

Land Trust Alliance. 2005. About land trusts. http://www.lta.org/aboutlt/index.html.

Landres, P. B., J. Verner, and J. W. Thomas. 1988. Critique of vertebrate indicator species. *Conservation Biology* 2:316–328.

Lashof, D. A., B. J. DeAngelo, S. R. Salesk, and J. Harte. 1997. Terrestrial ecosystem feedbacks to global climate change. *Annual Review of Energy and the Environment* 22:75–118.

Laurance, S. G. W. 2004. Responses of understory rain forest birds to road edges in central Amazonia. *Ecological Applications* 14:1344–1357.

Laurance, S. G., and W. F. Laurance. 1999. Tropical wildlife corridors: Use of linear rainforest remnants by arboreal mammals. *Biological Conservation* 91:231–239.

Laurance, S. G. W., P. C. Stouffer, and W. F. Laurance. 2004. Effects of road clearings on movement patterns of understory rainforest birds in central Amazonia. *Conservation Biology* 18:1099–1109.

Laurance, W. F. 1990. Comparative responses of five arboreal marsupials to tropical forest fragmentation. *Journal of Mammalogy* 71:641–653.

Laurance, W. F. 1991. Ecological correlates of extinction proneness in Australian tropical rainforest mammals. *Conservation Biology* 5:79–89.

Laurance, W. F. 1995. Rainforest mammals in a fragmented landscape. In *Landscape approaches in mammalian ecology and conservation*, ed. W. Z. Lidicker Jr., pages 46–63. Minneapolis: University of Minnesota Press.

Laurance, W. F. 1997. Hyper-disturbed parks: Edge effects and the ecology of isolated rainforest reserves in tropical Australia. In *Tropical forest remnants: Ecology, management, and conservation of fragmented communities*, eds. W. F. Laurance and R. O. J. Bierregaard Jr., pages 71–83. Chicago: University of Chicago Press.

Laurance, W. F., R. O. Bierregaard Jr., C. Gascon, R. K. Didham, A. P. Smith, A. J. Lynam, V. M. Viana et al. 1997. Tropical forest fragmentation: Synthesis of a diverse and dynamic discipline. In *Tropical forest remnants: Ecology, management, and conservation of fragmented communities*, eds. W. F. Laurance and R. O. Bierregaard Jr., pages 502–514. Chicago: University of Chicago Press.

Laurance, W. F., T. E. Lovejoy, H. L. Vasconcelos, E. M. Bruna, R. K. Didham, P. C. Stouffer, C. Gascon, R. O. Bierregaard, S. G. Laurance, and E. Sampaio. 2002. Ecosystem decay of Amazonian forest fragments: A 22-year investigation. *Conservation Biology* 16:605–618.

Le Maitre, D. C., B. W. van Wilgen, R. A. Chapman, and D. H. McKelly. 1996. Invasive plants and water resources in the Western Cape Province, South Africa: Modelling the consequences of a lack of management. *Journal of Applied Ecology* 33:161–172.

Leopold, A. 1933. *Game management.* New York: Chas. Scribner's Sons.

Levey, D. J., B. M. Bolker, J. J. Tewksbury, S. Sargent, and N. M. Haddad. 2005. Effects of landscape corridors on seed dispersal by birds. *Science* 309:146–148.

Levins, R. A. 1969. Some demographic and genetic consequences of environmental heterogeneity for biological control. *Bulletin of the Entomological Society of America* 15:237–240.

Levins, R. A. 1970. Extinction. *Lectures in Mathematical Life Sciences* 2:75–107.

Lichstein, J. W., T. R. Simons, and K. E. Franzreb. 2002. Landscape effects on breeding songbird abundance in managed forests. *Ecological Applications* 12:836–857.

Lidicker, W. Z., Jr. 1962. Emigration as a possible mechanism permitting the regulation of population density below carrying capacity. *American Naturalist* 96:29–33.

Lidicker, W. Z., Jr. 1975. The role of dispersal in the demography of small mammals. In *Small mammals: Their production and population dynamics*, eds. F. B. Golley, K. Petrusewicz, and L. Ryszkowski, pages 103–128. London: Cambridge University Press.

Lidicker, W. Z., Jr. 1978. Regulation of numbers in small mammal populations: Historical reflections and a synthesis. In *Populations of small mammals under natural conditions*, ed. D. P. Snyder, pages 122–141. Pittsburgh: University of Pittsburgh, Pymatuning Laboratory of Ecology, Special Publications, vol. 5.

Lidicker, W. Z., Jr. 1985. Population structuring as a factor in understanding microtine cycles. *Acta Zoologica Fennica* 173:23–27.

Lidicker, W. Z., Jr. 1988a. Solving the enigma of microtine "cycles." *Journal of Mammalogy* 69:225–235.

Lidicker, W. Z., Jr. 1988b. The synergistic effects of reductionist and holistic approaches in animal ecology. *Oikos* 53:278–281.

Lidicker, W. Z., Jr. 1989. Impacts of non-domesticated vertebrates on California grasslands. In *Grassland structure and function: California annual grassland*, eds. L. F. Huenneke and H. A. Mooney, pages 135–150. Dordrecht, Netherlands: Kluwer Academic Publishers.

Lidicker, W. Z., Jr. 1992. In defense of a multifactor perspective in population ecology. *Journal of Mammalogy* 72:631–635.

Lidicker, W. Z., Jr. ed. 1994. A spatially explicit approach to vole population processes. *Polish Ecological Studies* 20:215–225.

Lidicker, W. Z., Jr. 1995a. *Landscape approaches in mammalian ecology and conservation*. Minneapolis: University of Minnesota Press.

Lidicker, W. Z., Jr. 1995b. The landscape concept: Something old, something new. In *Landscape approaches in mammalian ecology and conservation*, ed. W. Z. Lidicker Jr., pages 3–19. Minneapolis: University of Minnesota Press.

Lidicker, W. Z., Jr. 1999. Responses of mammals to habitat edges: An overview. *Landscape Ecology* 14:333–343.

Lidicker, W. Z., Jr. 2000. A food web/landscape interaction model for microtine rodent density cycles. *Oikos* 91:435–445.

Lidicker, W. Z., Jr. 2002. From dispersal to landscapes: Progress in our understanding of population dynamics. *Acta Theriologica* 47 (Suppl. no. 1):23–37.

Lidicker, W. Z., Jr., and W. D. Koenig. 1996. Responses of terrestrial vertebrates to habitat edges and corridors. In *Metapopulations and wildlife conservation*, ed. D. R. McCullough, pages 85–109. Washington, DC: Island Press.

Lidicker, W. Z., Jr., and J. A. Peterson. 1999. Responses of small mammals to habitat edges. In *Landscape ecology of small mammals*, eds. G. W. Barrett and J. D. Peles, pages 211–227. New York: Springer-Verlag.

Lidicker, W. Z., Jr., and N. C. Stenseth. 1992. To disperse or not to disperse: Who does it and why? In *Animal dispersal: Small mammals as a model*, eds. N. C. Stenseth and W. Z. Lidicker Jr., pages 21–36. London: Chapman and Hall.

Lidicker, W. Z., Jr., J. O. Wolff, L. N. Lidicker, and M. H. Smith. 1992. Utilization of a habitat mosaic by cotton rats during a population decline. *Landscape Ecology* 6:259–268.

Lindborg, R., and O. Eriksson. 2004. Historical landscape connectivity affects present plant species diversity. *Ecology* 85:1840–1845.

Lindenmayer, D. B., and J. F. Franklin. 2002. *Conserving forest biodiversity: A comparative multiscaled approach*. Washington, DC: Island Press.

Liro, A., and J. Szacki. 1994. Movements of small mammals along two ecological corridors in suburban Warsaw. *Polish Ecological Studies* 20:227–231.

Lookingbill, T., and D. Urban. 2004. An empirical approach towards improved spatial estimates of soil moisture for vegetation analysis. *Landscape Ecology* 19:417–433.

Lorenz, G. C., and G. W. Barrett. 1990. Influence of simulated corridors on house mouse (*Mus musculus*) dispersal. *American Midland Naturalist* 12:348–356.

Lorenzana, J. C., and S. G. Sealy. 1999. A meta-analysis of the impact of parasitism by the brown-headed cowbird on its hosts. *Studies in Avian Biology* 18:241–253.

Loukmas, J. J., D. T. Mayack, and M. E. Richmond. 2003. Track plate enclosures: Box designs affecting attractiveness to riparian mammals. *American Midland Naturalist* 149:219–224.

Loveland, T. R., and H. L. Hutcheson. 1995. Monitoring changes in landscapes from satellite imagery. In *Our living resources: A report to the nation on the distribution, abundance, and health of U.S. plants, animals, and ecosystems*, eds. E. T. LaRoe, G. S. Farris, C. E. Puckett, P. D. Doran, and M. J. Mac, pages 468–473. Washington, DC: U.S. Department of the Interior, National Biological Service.

Lowrance, R., R. Todd, J. J. Fail, O. J. Hendrickson, R. Leonard, and L. Asmussen. 1984. Riparian forests as nutrient filters in agricultural watersheds. *BioScience* 34:374–377.

Lucy, W. H. 2003. Mortality risk associated with leaving home: Recognizing the relevance of the built environment. *American Journal of Public Health* 93:1564–1569.

Luers, M. 2004. Making your fence wildlife friendly. *Crossing paths with wildlife in Washington towns and cities* (newsletter of the Washington Department of Fish and Wildlife, Spokane), Fall.

Lynx News. 1998. Scientists scratch their heads over missing lynx population. November 15. http://lynx.uio.no/lynx/nancy/news/ny_top.htm.

MacArthur, R. H., and E. O. Wilson. 1967. *The theory of island biogeography*. Princeton, NJ: Princeton University Press.

Macdonald, D. W., and S. Rushton. 2003. Modelling space use and dispersal of mammals in real landscapes: A tool for conservation. *Journal of Biogeography* 30:607–620.

Machtans, C. S., M. A. Villard, and S. J. Hannon. 1996. Use of riparian buffer strips as movement corridors by forest birds. *Conservation Biology* 10:1366–1379.

Mack, M. C., and C. M. D'Antonio. 1998. Impacts of biological invasions on disturbance regimes. *Trends in Ecology and Evolution* 13:195–198.

Mac Nally, R., and A. F. Bennett. 1997. Species-specific prediction of the impact of habitat fragmentation: Local extinction of birds in the box-ironbark forests of central Victoria, Australia. *Biological Conservation* 82:147–155.

Maestas, J. D., R. L. Knight, and W. C. Gilgert. 2003. Biodiversity across a rural land-use gradient. *Conservation Biology* 17:1425–1434.

Malo, J. E., F. Suárez, and A. Díez. 2004. Can we mitigate animal-vehicle accidents using predictive models? *Journal of Applied Ecology* 41:701–710.

Margules, C. R., and R. L. Pressey. 2000. Systematic conservation planning. *Nature* 405:243–253.

Martin, K., P. B. Stacey, and C. E. Braun. 2000. Recruitment, dispersal, and demographic rescue in spatially-structured white-tailed ptarmigan populations. *Condor* 102:503–516.

Massolo, A., and A. Meriggi. 1998. Factors affecting habitat occupancy by wolves in northern Apennine (northern Italy): A model of habitat suitability. *Ecography* 21:97–107.

Mayer, K. E., and W. F. Laudenslayer Jr., eds. 1988. *A guide to wildlife habitats of California*. Sacramento: California Department of Fish and Game.

McCallum, H., and A. Dobson. 2002. Disease, habitat fragmentation and conservation. *Proceedings of the Royal Society of London, Series B: Biological Sciences* 269:2041–2049.

McCarthy, M. A., S. J. Andelman, and H. P. Possingham. 2003. Reliability of relative predictions in population viability analysis. *Conservation Biology* 17:982–989.

McCauley, D. E. 1993. Genetic consequences of extinction and recolonization in fragmented habitats. In *Biotic interactions and global change*, eds. P. M. Kareiva, J. G. Kingsolver, and R. B. Huey, pages 217–233. Sunderland, MA: Sinauer.

McCaull, J. 1994. The natural community conservation planning program and the coastal sage scrub ecosystem of Southern California. In *Environmental policy and biodiversity*, ed. R. E. Grumbine, pages 281–292. Washington, DC: Island Press.

McCloskey, J. M., and H. Spalding. 1989. A reconnaissance-level inventory of the amount of wilderness remaining in the world. *Ambio* 18:221–227.

McCullough, D. R., ed. 1996. *Metapopulations and wildlife conservation*. Washington, DC: Island Press.

McCullough, D. R., J. K. Fischer, and J. D. Ballou. 1996. From bottleneck to metapopulation: Recovery of the tule elk in California. pages 375–403 In ed. D. R. McCullough, *Metapopulations and wildlife conservation*. Washington, DC: Island Press.

McDonald, T. 2004. The Wilderness Society's Wild Country Program. *Ecological Management and Restoration* 5:87–97.

McDonald, W., and C. C. St. Clair. 2004. Elements that promote highway crossing structure use by small mammals in Banff National Park. *Journal of Applied Ecology* 41:82–93.

McGuffie, K., and A. Henderson-Sellers. 2004. Stable water isotope characterization of human and natural impacts on land-atmosphere exchanges in the Amazon Basin. *Journal of Geophysical Research* 109:1–25.

McKinney, M. L. 2002. Effects of national conservation spending and amount of protected area on species threat rates. *Conservation Biology* 16:539–543.

McMillion, S. 2004. Bears in town seeking food. *Bozeman (Montana) Daily Chronicle*, February 15, 2004.

McShea, W. J., H. B. Underwood, and J. H. Rappole, eds. 1997. *The science of overabundance: Deer ecology and population management*. Washington, DC: Smithsonian Institution Press.

Meadows, R. 2002. Scat sniffing. *Zoogoer* 31:5.

Meir, E., S. Andelman, and H. P. Possingham. 2004. Does conservation planning matter in a dynamic and uncertain world? *Ecology Letters* 7:615–622.

Merenlender, A. M. 2000. Mapping vineyard expansion provides information on agriculture and the environment. *California Agriculture* 54:7–12.

Merenlender, A. M., L. Huntsinger, G. Guthey, and S. K. Fairfax. 2004. Land trusts and conservation easements: Who is conserving what for whom? *Conservation Biology* 18:65–75.

Merenlender, A. M., C. Kremen, M. Rakotondratsima, and A. Weiss. 1998. Monitoring impacts of natural resource extraction on lemurs of the Masoala Peninsula, Madagascar. *Conservation Ecology* 2:5. http://www.consecol.org/.

Merenlender, A. M., K. L. Heise, C. Brooks. 1998. Effects of subdividing private property on biodiversity in California's north coast oak woodlands. *Transactions of the Wildlife Society* 34:9–20.

Merriam, G., and A. Lanoue. 1990. Corridor use by small mammals: Field measurement for three experimental types of *Peromyscus leucopus*. *Landscape Ecology* 4:123–131.

Mesquita, R. C. G., P. Delamônica, and W. F. Laurance. 1999. Effect of surrounding vegetation on edge-related tree mortality in Amazonian forest fragments. *Biological Conservation* 91:129–134.

Mietz, S. N. 1994. Linkage zone identification and evaluation of management options for grizzly bears in the Evaro Hill area. Ph.D. thesis, University of Montana, Missoula.

Miller, S. G., R. L. Knight, and C. K. Miller. 2001. Wildlife responses to pedestrians and dogs. *Wildlife Society Bulletin* 29:124–132.

Mills, L. S. 1996. Fragmentation of a natural area: Dynamics of isolation for small mammals on forest remnants. In *National parks and protected areas: Their role in environmental protection*, ed. G. Wright, pages 199–218. Cambridge, England: Blackwell Science.

Mills, L. S., and F. W. Allendorf. 1996. The one-migrant-per-generation rule in conservation and management. *Conservation Biology* 6:1509–1518.

Mills, L. S., and M. S. Lindberg. 2002. Sensitivity analysis to evaluate the consequences of conservation actions. pages 338–366 In eds. S. R. Beissinger and D. R. McCullough, *Population viability analysis*. Chicago: University of Chicago Press.

Mooney, K. A. 2002. Quantifying avian habitat use in forests using track-plates. *Journal of Field Ornithology* 73:392–398.

Mooney, H. A., and J. A. Drake, eds. 1986. *Ecology of biological invasions of North America and Hawaii*. New York: Springer-Verlag.

Morrison, R. I. G., R. K. Ross, and L. J. Niles. 2004. Declines in wintering populations of red knots in southern South America. *Condor* 106:60–70.

Mumme, R. L. 1994. Demographic consequences of roadside mortality in the Florida scrub jay. Abstract. *American Ornithologists' Union Annual Meeting*, Missoula, Montana.

Mumme, R. L., S. J. Schoech, G. W. Woolfenden, and J. W. Fitzpatrick. 2000. Life and death in the fast lane: Demographic consequences of road mortality in the Florida scrub jay. *Conservation Biology* 14:501–512.

Murcia, C. 1995. Edge effects in fragmented forests: Implications for conservation. *Trends in Ecology and Evolution* 10:58–62.

Murphy, E. 2005. Caught in the headlights. *High Country News*, February 7: 8–11.

Murphy, M. T. 2001. Source-sink dynamics of a declining eastern kingbird population and the value of sink habitats. *Conservation Biology* 15:737–748.

Naeem, S. 1998. Species redundancy and ecosystem reliability. *Conservation Biology* 12:39–45.

Naeem, S., L. J. Thompson, S. P. Lawler, J. H. Lawton, and R. M. Woodfin. 1994. Declining biodiversity can alter the performance of ecosystems. *Nature* 368:734–737.

National Resources Conservation Service. 2002. *Annual natural resource inventory: Land use*. Washington, DC: U.S. Department of Agriculture, Natural Resources and Conservation Services. http://www.nrcs.usda.gov/technical/land/nri02/nri02lu.html.

Natural Resources Conservation District. 1999. Conservation corridor planning at the landscape level: Managing for wildlife. *National biology handbook*. Washington, DC: U.S. Department of Agriculture, Natural Resources Conservation Service.

Ndubisi, F., T. Demeo, and N. D. Ditto. 1995. Environmentally sensitive areas: A template for developing greenway corridors. *Landscape and Urban Planning* 33:159–177.

Newburn, D., P. Berck, and A. M. Merenlender. 2006. Habitat and open space at risk of land-use conversion: Targeting strategies for land conservation. *American Journal of Agricultural Economics* 88:28–42.

Newburn, D., S. Reed, P. Berck, and A. M. Merenlender. 2005. Economics and land-use change in prioritizing private land conservation. *Conservation Biology* 19:1411–1420.

Newmark, W. D. 1987. A land-bridge island perspective on mammalian extinctions in western North America. *Nature* 325:430–432.

Newmark, W. D. 1995. Extinction of mammal populations in Western North-American national parks. *Conservation Biology* 9:512–526.

Ng, S. J., J. W. Dole, R. M. Sauvajot, S. P. D. Riley, and T. J. Valone. 2004. Use of highway undercrossings by wildlife in Southern California. *Biological Conservation* 115:499–507.

Nicholls, C. I., M. Parrella, and M. A. Altieri. 2001. The effects of a vegetational corridor on the abundance and dispersal of insect biodiversity within a northern California organic vineyard. *Landscape Ecology* 16:133–146.

Norton, D. A., R. J. Hobbs, and L. Atkins. 1995. Fragmentation, disturbance, and plant distribution: Mistletoes in woodland remnants in the Western Australian wheatbelt. *Conservation Biology* 9:426–438.

Norton, T. W., and H. A. Nix. 1991. Application of biological modelling and GIS to identify regional wildlife corridors. In *Nature conservation 2: The role of corridors*, eds. D. A. Saunders and R. J. Hobbs, pages 19–26. Chipping Norton, New South Wales, Australia: Surrey Beatty & Sons.

Noss, R. F. 1987. Corridors in real landscapes: A reply to Simberloff and Cox. *Conservation Biology* 1:159–164.

Noss, R. F. 1990. Indicators for monitoring biodiversity: A hierarchical approach. *Conservation Biology* 4:355–364.

Noss, R. F. 1991. Landscape connectivity: Different functions at different scales. In *Landscape linkages and biodiversity*, ed. W. E. Hudson, pages 27–39. Washington, DC: Island Press.

Noss, R. F. 2003. A checklist for wildlands network designs. *Conservation Biology* 17:1270–1275.

Noss, R. F., and B. Csuti. 1997. Habitat fragmentation. In *Principles of conservation biology*, 2nd ed., eds. G. K. Meffe and C. R. Carroll, pages 269–304. Sunderland, MA: Sinauer.

Nupp, T. E., and R. K. Swihart. 2000. Landscape-level correlates of small mammal assemblages in forest fragments of farmland. *Journal of Mammalogy* 81:512–526.

Odell, E. A., and R. L. Knight. 2001. Songbird and medium-sized mammal communities associated with exurban development in Pitkin County, Colorado. *Conservation Biology* 15:1143–1150.

Odell, E. A., T. M. Theobald, and R. L. Knight. 2003. Incorporating ecology into land use planning: The songbird's case for clustered development. *Journal of American Planning Association* 69:72–82.

Oksanen, T., L. Oksanen, and M. Gyllenberg. 1992. Exploitation ecosystems in heterogeneous habitat complexes II: Impact of small-scale heterogeneity on predator-prey dynamics. *Evolutionary Ecology* 6:383–398.

Oksanen, T., and M. Schneider. 1995. The influence of habitat heterogeneity on predator-prey dynamics. In *Landscape approaches in mammalian ecology and conservation*, ed. W. Z. Lidicker Jr., pages 122–150. Minneapolis: University of Minnesota Press.

Opdam, P., R. Foppen, R. Reijnen, and A. Schotman. 1995. The landscape ecological approach in bird conservation: Integrating the metapopulation concept into spatial planning. *Ibis* 137:139–146.

Opdam, P., and D. Wascher. 2004. Climate change meets habitat fragmentation: Linking landscape and biogeographical scale levels in research and conservation. *Biological Conservation* 117:285–297.

Opperman, J. J., and A. M. Merenlender. 2000. Deer herbivory as an ecological constraint to restoration of degraded riparian corridors. *Restoration Ecology* 8:41–47.

Orrock, J. L., and E. I. Damshen. 2005. Corridors cause differential seed predation. *Ecological Applications* 15:793–798.

Ortega, Y. K., and D. E. Capen. 1999. Effect of forest roads on habitat quality for ovenbirds in a forested landscape. *Auk* 116:937–946.

Ostfeld, R. S. 1992. Small-mammal herbivores in a patchy environment: Individual strategies and population responses. In *Effects of resource distribution on animal-plant interactions*, eds. M. D. Hunter, T. Ohgushi, and P. W. Price, pages 43–74. New York: Academic Press.

Ostlie, W. R., R. E. Schneider, J. M. Aldrich, T. M. Faust, R. L. B. McKim, and S. J. Chaplin. 1997. *The status of biodiversity in the Great Plains*. Arlington, VA: The Nature Conservancy.

Oza, G. M. 1983. Deteriorating habitat and prospects of the Asiatic lion. *Environmental Conservation* 10:349–352.

Palmquist, R. B. 1991. Hedonic methods. In *Measuring the demand for environmental quality*, eds. J. B. Braden and C. D. Kolstad, pages 77–120. Amsterdam: North-Holland.

Panetta, F. D. 1991. Negative values of corridors. In *Nature conservation 2: The role of corridors*, eds. D. A. Saunders and R. J. Hobbs, page 410. Chipping Norton, New South Wales, Australia: Surrey Beatty & Sons.

Panetta, F. D., and A. J. M. Hopkins. 1991. Weeds in corridors: Invasion and management. In *Nature conservation 2: The role of corridors*, eds. D. A. Saunders and R. J. Hobbs, pages 341–352. Chipping Norton, New South Wales, Australia: Surrey Beatty & Sons.

Papouchis, C. M., F. J. Singer, and W. B. Sloan. 2001. Responses of desert bighorn sheep to increased human recreation. *Journal of Wildlife Management* 65:573–582.

Paton, P. W. C. 1994. The effect of edge on avian nest success: How strong is the evidence? *Conservation Biology* 8:17–26.

Pearson, D. L., and F. Cassola. 1992. World-wide species richness patterns of tiger beetles (Coleoptera: Cicindelidae): Indicator taxon for biodiversity and conservation studies. *Conservation Biology* 6:376–391.

Peer, B. D., and S. G. Sealy. 2004. Correlates of egg rejection in hosts of the brown-headed cowbird. *Condor* 106:580–599.

Peluso, N. L. 1993. Coercing conservation: The politics of state resource control. *Global Environmental Change—Human and Policy Dimensions* 3:199–217.

Pence, G. Q. K., M. A. Botha, and J. K. Turpie. 2003. Evaluating combinations of on- and off-reserve conservation strategies for the Agulhas Plain, South Africa: A financial perspective. *Biological Conservation* 112:253–273.

Perault, D. R., and M. V. Lomolino. 2000. Corridors and mammal community structure across a fragmented, old-growth forest landscape. *Ecological Monographs* 70:401–422.

Peterson, J. A. 1996. Gray-tailed vole population responses to inbreeding and environmental stress. PhD diss., University of California, Berkeley.

Pimentel, D. 2002. *Biological invasions: Economic and environmental costs of alien plant, animal, and microbe species*. Boca Raton, FL: CRC Press.

Pimentel, D., L. Lach, R. Zuniga, and D. Morrison. 2000. Environmental and economic costs associated with non-indigenous species in the United States. *BioScience* 50:53–65.

Pletcher, S. D. 2004. Vital connections. *Science* 304:1570–1571.

Plotnick, R. E., and M. L. McKinney. 1993. Ecosystem organization and extinction dynamics. *Palaios* 8:202–212.

Poague, K. L., Johnson, R. J., and L. J. Young. 2000. Bird use of rural and urban converted railroad rights-of-way in southeast Nebraska. *Wildlife Society Bulletin* 28:852–864.

Posillico, M., M. I. B. Alberto, E. Pagnin, S. Lovari, and L. Russo. 2004. A habitat model for brown bear conservation and land use planning in the central Apennines. *Biological Conservation* 118:141–150.

Possingham, H. P., and I. Davies. 1995. ALEX: A model for the viability analysis of spatially structured populations. *Biological Conservation* 73:143–150.

Poulsen, J. R., and C. J. Clark. 2004. Densities, distributions, and seasonal movements of gorillas and chimpanzees in swamp forest in northern Congo. *International Journal of Primatology* 25:285–306.

Power, M. E., D. Tilman, J. A. Estes, B. A. Menge, W. J. Bond, L. S. Mills, G. Daily, J. C. Castilla, J. Lubchenco, and R. T. Paine. 1996. Challenges in the quest for keystones. *BioScience* 46:609–620.

Pressey, R., and K. Taffs. 2001. Scheduling conservation action in production landscapes: Priority areas in western New South Wales defined by irreplaceability and vulnerability to vegetation loss. *Biological Conservation* 100:355–376.

Price, O. F., J. C. Z. Woinarski, and D. Robinson. 1999. Very large area requirements for frugivorous birds in monsoon rainforests of the Northern Territory, Australia. *Biological Conservation* 91:169–180.

Primm, S. A., and T. W. Clark. 1996. Making sense of the policy process for carnivore conservation. *Conservation Biology* 10:1036–1045.

Pulliam, H. R. 1988. Sources, sinks, and population regulation. *American Naturalist* 132:652–661.

Pulliam, H. R., and B. J. Danielson. 1991. Sources, sinks, and habitat selection: A landscape perspective on population dynamics. *American Naturalist* (Suppl.) 137: S50–S66.

Puth, L. M., and K. A. Wilson. 2001. Boundaries and corridors as a continuum of ecological flow control: Lessons from rivers and streams. *Conservation Biology* 15:21–30.

Putman, R. J. 1997. Deer and road traffic accidents: Options for management. *Journal of Environmental Management* 51:43–57.

Queensland Parks and Wildlife Service. 2003. Wildlife corridors. http://www.epa.qld. gov.au/nature_conservation/community_role/landholders/case_studies/wildlife_ corridors.

Rail, J. F., M. Darveau, A. Desrochers, and J. Huot. 1997. Territorial responses of boreal forest birds to habitat gaps. *Condor* 99:976–980.

Rapp, V. 1997. *What the river reveals: Understanding and restoring healthy watersheds.* Seattle: Mountaineers Books.

Rebelo, A. G., and W. R. Siegfried. 1992. Where should nature-reserves be located in the Cape Floristic region, South Africa? Models for the spatial configuration of a reserve network aimed at maximizing the protection of floral diversity. *Conservation Biology* 6:243–252.

Redland Shire Council. 2002. *Fauna friendly fencing.* Fact sheet FS078. Cleveland, Australia: Author.

Reed, D. H. 2005. Relationship between population size and fitness. *Conservation Biology* 19:563–568.

Reijnen, R., and R. Foppen. 1994. The effects of car traffic on breeding bird populations in woodland. 1. Evidence of reduced habitat quality for willow warblers (*Phylloscopus trochilus*) breeding close to a highway. *Journal of Applied Ecology* 31:85–94.

Rejmánek, M., and D. M. Richardson. 1996. What attributes make some plant species more invasive? *Ecology* 77:1655–1661.

Reunanen, P., M. Mönkkönen, and A. Nikula. 2000. Managing boreal forest landscapes for flying squirrels. *Conservation Biology* 14:218–226.

Rhymer, J. M., and D. Simberloff. 1996. Extinction by hybridization and introgression. *Annual Review of Ecology and Systematics* 27:83–109.

Ricketts, T. H. 2004. Tropical forest fragments enhance pollinator activity in nearby coffee crops. *Conservation Biology* 18:1262–1271.

Ricketts, T. H., G. C. Daily, P. R. Ehrlich, and J. P. Fay. 2001. Countryside biogeography of moths in a fragmented landscape: Species richness in native and agricultural habitats. *Conservation Biology* 15:378–389.

Ricketts, T. H., G. C. Daily, P. R. Ehrlich, and C. D. Michener. 2004. Economic value of tropical forest to coffee production. *Proceedings of the National Academy of Sciences of the United States of America* 101:12579–12582.

Ries, L., R. J. Fletcher Jr., J. Battin, and T. D. Sisk. 2004. Ecological responses to habitat edges: Mechanisms, models, and variability explained. *Annual Review of Ecology Evolution and Systematics* 35:491–522.

Ripple, W. J., and R. L. Beschta. 2004. Wolves, elk, willows, and trophic cascades in the upper Gallatin Range of southwestern Montana, USA. *Forest Ecology and Management* 200:161–181.

Roberge, J. M., and P. Angelstam. 2004. Usefulness of the umbrella species concept as a conservation tool. *Conservation Biology* 18:76–85.

Robinson, S. K., S. I. Rothstein, M. C. Brittingham, L. J. Petit, and J. A. Grzybowski. 1995. Ecology and behavior of cowbirds and their impact on host populations. In *Ecology and management of neotropical migratory birds*, eds. T. E. Martin and D. M. Finch, pages 428–460. New York: Oxford University Press.

Robinson, S. K., F. R. Thompson III, T. M. Donovan, D. R. Whitehead, and J. Faaborg. 1995. Regional forest fragmentation and the nesting success of migratory birds. *Science* 267:1987–1990.

Rodenhouse, N. L., G. W. Barreto, D. M. Zimmerman, and J. C. Kemp. 1992. Effects of uncultivated corridors on arthropod abundances and crop yields in soybean agroecosystem. *Agriculture Ecosystems and Environment* 38:179–191.

Rodrigues, A. S. L., S. J. Andelman, M. I. Bakarr, L. Boitani, T. M. Brooks, R. M. Cowling, L. D. C. Fishpool et al. 2004. Effectiveness of the global protected area network in representing species diversity. *Nature* 428:640–643.

Rodriguez, A., G. Crema, and M. Delibes. 1996. Use of non-wildlife passage across a high speed railway by terrestrial vertebrates. *Journal of Applied Ecology* 33:1527–1540.

Roff, D. 2003. Evolutionary danger for rainforest species. *Science* 301:58–59.

Roll, J., R. J. Mitchell, R. J. Cabin, and D. L. Marshall. 1997. Reproductive success increases with local density of conspecifics in a desert mustard (*Lesquerella fendleri*). *Conservation Biology* 11:738–746.

Rolstad, J. 1991. Consequences of forest fragmentation for the dynamics of bird populations: Conceptual issues and the evidence. *Journal of the Linnean Society* 42:149–163.

Rolston, H. 1988. *Environmental ethics: Duties to and values in the natural world.* Philadelphia: Temple University Press.

Rosen, S. 1974. Hedonic prices and implicit markets: Product differentiation in pure competition. *Journal of Political Economy* 82:34–55.

Rosenberg, D. K., B. R. Noon, and E. C. Meslow. 1997. Biological corridors: Form, function, and efficacy. *BioScience* 47:677–687.

Rosenblatt, D. L., E. J. Heske, S. L. Nelson, D. M. Barber, M. A. Miller, and B. Macallister. 1999. Forest fragments in east-central Illinois: Islands or habitat patches for mammals? *American Midland Naturalist* 141:115–123.

Rothstein, S. I. 1994. The cowbird's invasion of the Far West: History, causes and consequences experienced by host species. *Studies in Avian Biology* 15:301–315.

Royte, E. 2002. Wilding America: Connect our last parcels of wilderness, like pearls on a necklace, and mountain lions, bobcats, and wolves might once again roam their ancestral ranges. *Discover* 23, no. 9, http://www.discover.com/sept_02/featwild.html.

Ruxton, G. D. 1994. Low levels of immigration between chaotic populations can reduce system extinctions by inducing asynchronous regular cycles. *Proceedings of the Royal Society of London, Series B: Biological Sciences* 256:189–193.

Ryszkowski, L., N. R. French, and A. Kędziora, eds. 1996. *Dynamics of an agricultural landscape*. Poznań: Polish Academy of Sciences, Zakład Badań Środowiska Rolniczego i Leśnego [Research Center for Agricultural and Forest Environment].

Sakai, H. F., and B. R. Noon. 1997. Between-habitat movements of dusky-footed woodrats and vulnerability to predation. *Journal of Wildlife Management* 61:348–350.

Sala, O. E., F. S. Chapin, J. J. Armesto, E. Berlow, J. Bloomfield, R. Dirzo, E. Huber-Sanwald et al. 2000. Biodiversity: Global biodiversity scenarios for the year 2100. *Science* 287:1770–1774.

Samson, F., and F. Knopf. 1994. Prairie conservation in North America. *BioScience* 44:418–421.

Sanderson, E. W., M. Jaiteh, M. A. Levy, K. H. Redford, A. V. Wannebo, and G. Woolmer. 2002. The human footprint and the last of the wild. *BioScience* 52:891–904.

Sanderson, J. G., and L. D. Harris. 1999. *Landscape ecology: A top down approach*. Boca Raton, FL: Lewis Publishers.

Saraiva, M. G., G. M. Kondolf, P. Simoes, K. Morgado, and J. Alves. 2002. River conservation and restoration potential in the metropolitan area of Lisbon, Portugal. *Fourth International Workshop on Sustainable Land Use Planning: Collaborative planning for the metropolitan landscape*. Bellingham, WA: Western Washington University.

Saunders, D. A., and C. P. de Rebeira. 1991. Values of corridors to avian populations in a fragmented landscape. In *Nature conservation 2: The role of corridors*, eds. D. A. Saunders and R. J. Hobbs, pages 221–240. Chipping Norton, New South Wales, Australia: Surrey Beatty & Sons.

Saunders, D. A., and R. J. Hobbs, eds. 1991. *Nature conservation 2: The role of corridors*. Chipping Norton, New South Wales, Australia: Surrey Beatty & Sons.

Schlaepfer, M. A., M. C. Runge, and P. W. Sherman. 2002. Ecological and evolutionary traps. *Trends in Ecology and Evolution* 117:474–480.

Schlotterbeck, J. 2003. Preserving biological diversity with wildlife corridors: Amending the guidelines to the California Environmental Quality Act. *Ecology Law Quarterly* 30:955–990.

Schmidt, N. M., and P. M. Jensen. 2003. Changes in mammalian body length over 175 years: Adaptations to a fragmented landscape? *Conservation Ecology* 7:6. http://www.consecol.org/.

Schmiegelow, F. K. A., S. G. Cumming, S. Harrison, S. J. Leroux, K. A. Lisgo, R. F. Noss, and B. T. Olsen. 2006. Conservation beyond crisis management: A new model for the world's remaining intact areas. In prep.

Schmiegelow, F. K. A., and M. Mönkkönen. 2002. Habitat loss and fragmentation in dynamic landscapes: Avian perspective from the boreal forest. *Ecological Applications* 12:375–389.

Schultz, C. B. 1995. Corridors, islands and stepping stones: The role of dispersal behavior in designing reserves for a rare Oregon butterfly. *Bulletin of the Ecological Society of America* 76:240.

Schumaker, N. H., T. Ernst, D. White, J. Baker, and P. Haggerty. 2004. Projecting wildlife responses to alternative future landscapes in Oregon's Willamette Basin. *Ecological Applications* 14:381–400.

Scott, J. C. 1999. *Seeing like a state: How certain schemes to improve the human condition have failed.* New Haven, CT: Yale University Press.

Scott, J. M., F. W. Davis, R. G. McGhie, R. G. Wright, C. Groves, and J. Estes. 2001. Nature reserves: Do they capture the full range of America's biological diversity? *Ecological Applications* 11:999–1007.

Scott, M. 2002. *Predicting species occurrences: Issues of accuracy and scale.* Washington, DC: Island Press.

Seal, U. S., E. T. Thorne, M. A. Bogan, and S. H. Anderson. 1989. *Conservation biology and the black-footed ferret.* New Haven, CT: Yale University Press.

Sefa Dei, G. J. 2000. Rethinking the role of indigenous knowledge in the academy. *International Journal of Inclusive Education* 4:111–132.

Segurado, P., and M. B. Araujo. 2004. An evaluation of methods for modelling species distributions. *Journal of Biogeography* 31:1555–1568.

Selman, P. 1993. Landscape ecology and countryside planning: Vision, theory and practice. *Journal of Rural Studies* 9:1–21.

Selonen, V., and I. K. Hanski. 2003. Movements of the flying squirrel *Pteromys volans* in corridors and in matrix habitat. *Ecography* 26:641–651.

Semlitsch, R. D., and J. R. Bodie. 2003. Biological criteria for buffer zones around wetlands and riparian habitats for amphibians and reptiles. *Conservation Biology* 17:1219–1228.

Shaffer, M. L. 1981. Minimum population sizes for species conservation. *BioScience* 31:131–134.

Sheppe, W. 1972. The annual cycle of small mammal populations on a Zambian flood plain. *Journal of Mammalogy* 53:445–460.

Sieving, K. E., T. A. Contreras, and K. L. Maute. 2004. Heterospecific facilitation of forest-boundary crossing by mobbing understory birds in north-central Florida. *Auk* 121:738–751.

Sieving, K. E., M. F. Willson, and T. L. De Santo. 2000. Defining corridor functions for endemic birds in fragmented south-temperate rainforest. *Conservation Biology* 14:1120–1132.

Simberloff, D. 2001. Management of boreal forest biodiversity: A view from the outside. *Scandinavian Journal of Forest Research.* Suppl. no. 3: 105–118.

Simberloff, D. S., and L. G. Abele. 1976. Island biogeography theory and conservation practice. *Science* 191:285–286.

Simberloff, D., and J. Cox. 1987. Consequences and costs of conservation corridors. *Conservation Biology* 1:63–71.

Simberloff, D., J. A. Farr, J. Cox, and D. Mehlman. 1992. Movement corridors: Conservation bargains or poor investments. *Conservation Biology* 6:493–504.

Sizer, N., and E. V. J. Tanner. 1999. Responses of woody plant seedlings to edge formation in a lowland tropical rainforest, Amazonia. *Biological Conservation* 91:135–142.

Smallwood, K. S. 1994. Trends in California mountain lion populations. *Southwestern Naturalist* 39:7–72.

Smith, A. T., and M. Gilpin. 1997. Spatially correlated dynamics in a pika population. In *Metapopulation biology: Ecology, genetics, and evolution*, eds. I. A. Hanski and M. E. Gilpin, pages 407–428. San Diego, CA: Academic Press.

Smith, D. A., K. Ralls, B. Davenport, B. Adams, and J. E. Maldonado. 2001. Canine assistants for conservationists. *Science* 291:435.

Society for Ecological Restoration International Science and Policy Working Group. 2004. *The SER international primer on ecological restoration*. Tucson, AZ: Author.

Söndgerath, D., and B. Schröder. 2001. Population dynamics and habitat connectivity affecting the spatial spread of populations: A simulation study. *Landscape Ecology* 17:57–70.

Sonoma County Agricultural Preservation and Open Space District. 2000. *Acquisition plan: A blueprint for agricultural and open space preservation*. Santa Rosa, CA: Author.

Soulé, M. E. 1985. Biodiversity indicators in California: Taking nature's temperature. *California Agriculture* 49:40–44.

Soulé, M. E., and M. E. Gilpin. 1991. The theory of wildlife corridor capability. In *Nature conservation 2: The role of corridors*, eds. D. A. Saunders and R. J. Hobbs, pages 3–8. Chipping Norton, New South Wales, Australia: Surrey Beatty & Sons.

Soulé, M. E. and G. H. Orians, eds. 2001. *Conservation biology: Research priorities for the next decade*. Washington, DC: Island Press.

Soulé, M. E., and M. A. Sanjayan. 1998. Conservation targets: Do they help? *Science* 279:2060–2061.

Spackman, S. C., and J. W. Hughes. 1995. Assessment of minimum stream corridor width for biological conservation: Species richness and distribution along mid-order streams in Vermont, USA. *Biological Conservation* 71:325–332.

Spencer, W. D., R. H. Barrett, and W. J. Zielinski. 1983. Marten habitat preferences in the northern Sierra Nevada. *Journal of Wildlife Management* 47:1181–1186.

Stacey, P. B., M. L. Taper, and V. A. Johnson. 1997. Migration within metapopulations: The impact upon local population dynamics. pages 267–291 In *Metapopulation biology: Ecology, genetics and evolution*. eds. I. Hanski and M. E. Gilpin. San Diego, CA: Academic Pres

Stamps, J. A., M. Buechner, and V. V. Krishnan. 1987a. The effects of edge permeability and habitat geometry on emigration from patches of habitat. *American Naturalist* 129:533–552.

Stamps, J. A., M. Buechner, and V. V. Krishnan. 1987b. The effects of habitat geometry on territorial defense costs: Intruder pressure in bounded habitats. *American Zoologist* 27:307–325.

Standiford, R. B., and T. A. Scott. 2001. Value of oak woodlands and open space on private property values in Southern California. Special issue, *Investigación Agraria: Sistemas y Recursos Forestales* 1:137–152.

Stapp, P., M. F. Antolin, and M. Ball. 2004. Patterns of extinction in prairie dog metapopulations: Plague outbreaks follow El Niño events. *Frontiers in Ecology and the Environment* 2:235–240.

St. Clair, C. C., M. Bélisle, A. Desrochers, and S. Hannon. 1998. Winter responses of forest birds to habitat corridors and gaps. *Conservation Ecology* 2:13. http://www.consecol. org/.

Stefan, Å. 1999. Invasion of matrix species in small habitat patches. *Conservation Ecology* 3:1–14. http://www.consecol.org/.

Steffan-Dewenter, I. 2003. Importance of habitat area and landscape context for species richness of bees and wasps in fragmented orchard meadows. *Conservation Biology* 17:1036–1044.

Steidl, R. J., and R. G. Anthony. 1996. Responses of bald eagles to human activity during the summer in interior Alaska. *Ecological Applications* 6:482–491.

Stenseth, N. C., and W. Z. Lidicker Jr. 1992. The study of dispersal: A conceptual guide. In *Animal dispersal: Small mammals as a model.* eds. N. C. Stenseth and W. Z. Lidicker Jr. pages 5–20. London: Chapman and Hall.

Stephens, S. E., D. N. Koons, J. J. Rotella, and D. W. Willey. 2003. Effects of habitat fragmentation on avian nesting success: A review of the evidence at multiple spatial scales. *Biological Conservation* 115:101–110.

Sternberg, M. A. 2003. Comparison of native brushland, replanted, and unaided secondary succession plant communities in the Lower Rio Grand Valley of Texas. *Texas Journal of Science* 55:129–148.

Stith, B. M., J. W. Fitzpatrick, G. E. Woolfenden, and B. Pranty. 1996. Classification and conservation of metapopulations: A case study of the Florida scrub jay. In *Metapopulations and wildlife conservation*, ed. D. R. McCullough, pages 187–215. Washington, DC: Island Press.

Stoms, D. M., P. J. Comer, P. J. Crist, and D. H. Grossman. 2005. Choosing surrogates for biodiversity conservation in complex planning environments. *Journal of Conservation Planning* 1:44–63.

Stoms, D. M., F. W. Davis, and C. B. Cogan. 1992. Sensitivity of wildlife habitat models to uncertainties in GIS data. *Photogrammetric Engineering & Remote Sensing* 58:843–850.

Stromberg, J. C. 1997. Growth and survivorship of Fremont cottonwood, Gooding willow, and salt cedar seedling, after large floods in central Arizona. *Great Basin Naturalist* 57:198–208.

Stromberg, M. R., and J. R. Griffin. 1996. Long-term patterns in coastal California grasslands in relation to cultivation, gophers, and grazing. *Ecological Applications* 6:1189–1211.

Struyk, R., and K. Angelici. 1996. The Russian dacha phenomenon. *Housing Studies* 11:233–261.

Sutton, P., T. J. Cova, and C. D. Elvidge. 2006. Mapping "exurbia" in the conterminous United States using nighttime satellite imagery. *Geocarto.* In press.

Szacki, J., and A. Liro. 1991. Movements of small mammals in the heterogeneous landscape. *Landscape Ecology* 5:219–224.

Tabarelli, M., W. Mantovani, and C. A. Peres. 1999. Effects of habitat fragmentation on plant guild structure in the montane Atlantic forest of southeastern Brazil. *Biological Conservation* 91:119–127.

Tast, J. 1966. The root vole, *Microtus oeconomus* (Pallas), as an inhabitant of a seasonally flooded land. *Annals Zoologici Fennici* 3:127–171.

Taylor, A. D. 1991. Studying metapopulation effects in predator-prey systems. *Biological Journal of the Linnean Society* 42:305–323.

Taylor, A. R., and R. L. Knight. 2003. Wildlife responses to recreation and associated visitor perceptions. *Ecological Applications* 13:951–963.

Taylor, B. 1991. Investigating species incidence over habitat fragments of different areas: A look at error estimation. *Biological Journal of the Linnean Society* 42:177–191.

Taylor, P. D., L. Fahrig, K. Henein, and G. Merriam. 1993. Connectivity is a vital element of landscape structure. *Oikos* 68:571–573.

Temperton, V. M., R. J. Hobbs, T. Nuttle, and S. Halle, eds. 2004. *Assembly rules and restoration ecology: Bridging the gap between theory and practice.* Washington, DC: Island Press.

Templeton, A. R. 1986. Coadaptation and outbreeding depression. In *Conservation biology: The science of scarcity and diversity,* ed. M. E. Soulé, pages 105–116. Sunderland, MA: Sinauer.

Terborgh, J., L. Lope, P. Nuñez, M. Rao, G. Shahabuddin, G. Orihuela, M. Riverus et al. 2001. Ecological meltdown in predator-free forest fragments. *Science* 294:1923–1926.

Tewksbury, J. J., D. J. Levey, N. M. Haddad, S. Sargent, J. L. Orrock, A. Weldon, B. J. Danielson, J. Brinkerhoff, E. I. Damschen, and P. Townsend. 2002. *Proceedings of the National Academy of Sciences* 99:12923–12926.

Theobald, D. M. 2001. Land-use dynamics beyond the American urban fringes. *Geographical Review* 91:544–564.

Theobald, D. M., and N. T. Hobbs. 1998. Forecasting rural land use change: A comparison of regression- and spatial transition-based models. *Geographical and Environmental Modelling* 2:65–82.

Tilman, D. 1990. Constraints and tradeoffs: Toward a predictive theory of competition and succession. *Oikos* 58:3–15.

Tilman, D. 1994. Competition and biodiversity in spatially structured habitats. *Ecology* 75:2–16.

Tilman, D., and P. Kareiva, eds. 1997. *Spatial ecology: The role of space in population dynamics and interspecific interactions.* Princeton, NJ: Princeton University Press.

Tilman, D., C. L. Lehman, and P. Kareiva. 1997. Population dynamics in spatial habitats. In *Spatial ecology: The role of space in population dynamics and interspecific interactions,* eds. D. Tilman and P. Kareiva, pages 3–20. Princeton, NJ: Princeton University Press.

Tinker, D. B., W. H. Romme, and D. G. Despain. 2003. Historic range of variability in landscape structure in subalpine forests of the Greater Yellowstone Area, USA. *Landscape Ecology* 18:427–439.

Tischendorf, L., and L. Fahrig. 2000. On the usage and measurement of landscape connectivity. *Oikos* 90:7–19.

Tjallingii, S. P. 2000. Ecology on the edge: Landscape and ecology between town and country. *Landscape and Urban Planning* 48:103–119.

Trade and Environment. 1996. Costa Rica eco-tourism. American University, School of International Service, Washington, DC. http://www.american.edu/TED/costtour.htm.

Tress, B., and G. Tress. 2001. Capitalising on multiplicity: A transdisciplinary systems approach to landscape research. *Landscape and Urban Planning* 57:143–157.

Treves, A., L. Naughton-Treves, E. K. Harper, D. J. Mladenoff, R. A. Rose, T. A. Sickley, and A. P. Wydeven. 2004. Predicting wildlife-human conflict: A spatial model derived from 25 years of data on wolf predation on livestock. *Conservation Biology* 18:114–125.

Trzcinski, M. K., L. Fahrig, and G. Merriam. 1999. Independent effects of forest cover and fragmentation on the distribution of forest breeding birds. *Ecological Applications* 9:586–593.

Tucker, N. I. J. 2000. Linkage restoration: Interpreting fragmentation theory for the design of a rainforest linkage in the humid wet tropics of north-eastern Queensland. *Ecological Management and Restoration* 1:35–41.

Turchin, P. 2003. *Complex population dynamics: A theoretical/empirical synthesis.* Princeton, NJ: Princeton University Press.

Turner, M. G. 1989. Landscape ecology: The effect of pattern on process. *Annual Review of Ecology and Systematics* 20:171–197.

Turner, M. G., V. H. Dale, and R. H. Gardner. 1989. Predicting across scales: Theory development and testing. *Landscape Ecology* 3:245–252.

Tuxbury, S. M., and M. Salmon. 2005. Competitive interactions between artificial lighting and natural cues during sea finding by hatchling marine turtles. *Biological Conservation* 121:311–316.

Tyre, A. J., H. P. Possingham, and D. B. Lindenmayer. 2001. Inferring process from pattern: Can territory occupancy provide information about life history parameters? *Ecological Applications* 11:1722–1737.

United Republic of Tanzania. 1998. *The wildlife policy of Tanzania.* Dar es Salaam, Tanzania: Ministry of Natural Resources and Tourism.

Urquhart, D. 2004. Fingerprinting grizzlies: U of C PhD grad uses forensics to map bruin numbers in Alberta and BC. *OnCampus Weekly* (University of Calgary), January 23, 2004.

USDA/USDI. 1994. *Standards and guidelines for management of habitat for late-successional and old-growth forest related species within the range of the northern spotted owl.* Washington, DC: U.S. Department of Agriculture, Bureau of Land Management, and U.S. Department of Interior, Forest Service.

van Bohemen, H. D. 1996. Mitigation and compensation of habitat fragmentation caused by roads: Strategy, objectives, and practical measures. *Transportation Research Record no. 1475.* Delft, Netherlands: Ministry of Transport, Public Works, and Water Management.

van Bohemen, H. D. 2002. Infrastructure, ecology and art. *Landscape and Urban Planning* 59:187–201.

van Bohemen, H. D., and D. M. Teodorascu. 1997. Habitat fragmentation and de-fragmentation by motorways; mitigation and compensation principles; state-of-the-art report on European perspective. In *Infrastructuele Ontwikkeling.* Dutch Ministry of Transport, Public Works and Water Management.

van den Berg, L., and A. Wintjes. 2000. New rural "live style estates" in The Netherlands. *Landscape and Urban Planning* 48:169–176.

van der Ree, R., and A. F. Bennett. 2003. Home range of the squirrel glider (*Petaurus norfolcensis*) in a network of remnant linear habitats. *Journal of Zoology* 259:327–336.

van der Ree, R., A. F. Bennett, and D. C. Gilmore. 2003. Gap-crossing by gliding marsupials: Thresholds for use of isolated woodland patches in an agricultural landscape. *Biological Conservation* 115:241–249.

Vane-Wright, R. I., C. J. Humphries, and P. H. Williams. 1991. What to protect? Systematics and the agony of choice. *Biological Conservation* 55:235–254.

Van Horne, B. 1983. Density as a misleading indicator of habitat quality. *Journal of Wildlife Management* 47:893–901.

van Langevelde, F. 2000. Scale of habitat connectivity and colonization in fragmented nuthatch populations. *Ecography* 23:614–622.

Vannote, R. L., W. G. Minshall, K. W. Cummins, J. S. Sedell, and C. E. Cushing. 1980. The river continuum concept. *Canadian Journal of Fisheries and Aquatic Sciences* 37:130–137.

Van Valen, L. M. 1973. Pattern and the balance of nature. *Evolutionary Theory* 1:31–49.

Vitousek, P. M., C. M. D'Antonio, L. L. Loope, and R. Westbrooks. 1996. Biological invasions as global environmental change. *American Scientist* 84:468–478.

Vuilleumier, S., and R. Prelaz-Droux. 2002. Map of ecological networks for landscape planning. *Landscape and Urban Planning* 58:157–170.

Wade, T. G., K. H. Ritters, J. D. Wickham, and K. B. Jones. 2003. Distribution and causes of global forest fragmentation. *Conservation Ecology* 7:7. http://www.consecol.org/.

Walker, B. H. 1992. Biodiversity and ecological redundancy. *Conservation Biology* 6:18–23.

Walker, R., and L. Craighead. 1997. Analyzing wildlife movement corridors in Montana using GIS. *ESRI User Conference Proceedings, California. http://gis.esri.com/library/user-conf/proc97/proc97/abstract/a116.htm.*

Wallace, J. B., S. L. Eggert, J. L. Meyer, and J. R. Webster. 1997. Multiple trophic levels of a forest stream linked to terrestrial litter inputs. *Science* 277:102–104.

Wang, J. 2004. Application of the one-migrant-per-generation rule to conservation and management. *Conservation Biology* 18:332–343.

Wang, M. Y. L. 1997. The disappearing rural-urban boundary: Rural socioeconomic transformation in the Shenyang-Dalian region of China. *Third World Planning Review* 19:229–250.

Watchman, L. H., M. Groom, and J. D. Perrine. 2001. Science and uncertainty in habitat conservation planning. *American Scientist* 89:351–359.

Watson, J. R. 1991. The identification of river foreshore corridors for nature conservation in the south coast region of Western Australia. In *Nature conservation 2: The role of corridors*, eds. D. A. Saunders and R. J. Hobbs, pages 63–68. Chipping Norton, New South Wales, Australia: Surrey Beatty & Sons.

Weber, B. 2004. The arrogance of America's designer ark. *Conservation Biology* 18:1–3.

Wegner, J., and G. Merriam. 1990. Use of spatial elements in a farmland mosaic by woodland rodents. *Biological Conservation* 54:263–276.

Weiss, S. 1999. Cars, cows, and checkerspot butterflies: Nitrogen deposition and management of nutrient-poor grasslands for a threatened species. *Conservation Biology* 13:1476–1486.

Wells, M. L., J. F. O'Leary, J. Franklin, J. Michaelsen, and D. E. McKinsey. 2004. Variations in a regional fire regime related to vegetation type in San Diego County, California (USA). *Landscape Ecology* 19:139–152.

Westemeier, R. L., J. D. Brawn, S. A. Simpson, T. L. Esker, R. W. Jansen, J. W. Walk, E. L. Kershner, J. L. Bouzat, and K. N. Paige. 1998. Tracking the long-term decline and recovery of an isolated population. *Science* 282:1695–1698.

Wiens, J. A. 1985. Vertebrate responses to environmental patchiness in arid and semiarid ecosystems. In *The ecology of natural disturbance and patch dynamics*, eds. S. T. A. Pickett and P. S. White, pages 169–193. New York: Academic Press.

Wilcove, D., M. Bean, R. Bonnie, and M. McMillan. 1996. *Rebuilding the arc: Toward a more effective Endangered Species Act for private land.* Washington, DC: Environmental Defense Fund.

Wilcox, B. A. 1978. Supersaturated island faunas: A species-age relationship for lizards on post-Pleistocene land-bridge islands in the Gulf of California. *Science* 199:996–998.

Wildlife Trust. Brazil projects. http://www.wildlifetrust.org/brazil.htm.

Wiles, G. J., J. Bart, R. E. Beck, and C. F. Aguon. 2003. Impacts of the brown tree snake: Patterns of decline and species persistence in Guam's avifauna. *Conservation Biology* 17:1350–1360.

Wilhere, G. F. 2002. Adaptive management in habitat conservation plans. *Conservation Biology* 16:20–29.

Williams, J. C., C. S. ReVelle, and S. A. Levin. 2004. Using mathematical optimization models to design nature reserves. *Frontiers in Ecology and the Environment* 2:98–105.

Wilson, D. S. 1992. Complex interactions in metacommunities, with implications for biodiversity and higher levels of selection. *Ecology* 73:1984–2000.

Wilson, E. O. 1988. The current state of biological diversity. In *Biodiversity*, eds. E. O. Wilson and F. M. Peter, pages 3–18. Washington, DC: National Academy Press.

Wilson, K., R. Pressey, A. Newton, M. Burgman, H. P. Possingham, and C. Weston. 2005. Measuring and incorporating vulnerability into conservation planning. *Environmental Management* 35:527–543.

Wolff, J. O., and W. Z. Lidicker Jr. 1980. Population ecology of the taiga vole, *Microtus xanthognathus*, in interior Alaska. *Canadian Journal of Zoology* 58:1800–1812.

Wolff, J. O., E. M. Schauber, and W. D. Edge. 1996. Can dispersal barriers really be used to depict emigrating small mammals? *Canadian Journal of Zoology* 74:1826–1830.

Wolff, J. O., E. M. Schauber, and W. D. Edge. 1997. Effects of habitat loss and fragmentation on the behavior and demography of gray-tailed voles. *Conservation Biology* 11:945–956.

Wood, P. A., and M. J. Samways. 1991. Landscape element pattern and continuity of butterfly flight paths in an ecologically landscaped botanic garden, Natal, South Africa. *Biological Conservation* 58:149–166.

Wright, R. G., and P. D. Tanimoto. 1998. Using GIS to prioritize land conservation actions: Integrating factors of habitat diversity, land ownership, and development risk. *Natural Areas Journal* 18:38–44.

Yahner, R. H. 1988. Changes in wildlife communities near edges. *Conservation Biology* 2:333–339.

Zanette, L., P. Doyle, and S. M. Trémont. 2000. Food shortage in small fragments: Evidence from an area-sensitive passerine. *Ecology* 81:1654–1666.

Zanette, L., and B. Jenkins. 2000. Nesting success and nest predators in forest fragments: A study using real and artificial nests. *Auk* 117:445–454.

Zavaleta, E. S., and K. B. Hulvey. 2004. Realistic species losses disproportionately reduce grassland resistance to biological invaders. *Science* 306:1175–1177.

Zhou, G., and A. M. Liebhold. 1995. Forecasting the spatial dynamics of gypsy moth outbreaks using cellular transition models. *Landscape Ecology* 10:177–189.

Zube, E. H. 1995. Greenways and the US national park system. *Landscape and Urban Planning* 33:17–25.

About the Authors

Dr. Hilty is a landscape ecologist and conservation biologist with a research focus on understanding thresholds of human impact on biodiversity. Her experiences range from development of tools for community-based conservation efforts and enhancement of techniques to assess ecosystem health to developing new methodologies for assessing landscape connectivity and modeling of species occurrence and distribution across landscapes. In addition to writing a dissertation on wildlife corridors, she has direct experience working with communities on corridor design and protection issues. She is currently based in Bozeman, Montana, serving as Assistant Director for the Wildlife Conservation Society's North America program and helping to oversee a diverse program of conservation and research initiatives.

Dr. Lidicker is an ecologist, conservation biologist, and vertebrate zoologist. With over fifty years in academia, he brings extensive experience to bear on the subject of this book. His research career spans many disciplines, including population dynamics, social behavior, population genetics, mammalian systematics, evolution, and landscape ecology. During his tenure at the University of California Berkeley, he has published extensively and taught numerous courses encompassing those and other interests. He is currently Professor of Integrative Biology and Curator of Mammals (Museum of Vertebrate Zoology) emeritus.

Dr. Merenlender is a conservation biologist. In particular, she is interested in the forces that influence loss of biodiversity at all hierarchical levels, from genes to ecosystems. Her experience ranges from single-species management to measures of ecosystem health and most recently to regional land-use planning. She has published research papers on ecological monitoring, restoration ecology, cumulative impacts to watersheds, conservation easements, protected area planning, and invasive species. Through her position as a Cooperative Extension Specialist at the University of California Berkeley, she translates scientific findings into lay terms for policy makers, private landowners, and public land managers.

Index

Light effects. *See* Artificial lights
Linkage. *See* Corridor
Lion (*Leo leo persica*) protection, 24, 37
Lion, mountain (*Puma concolor*): corridors, 99–100, *102–103*, 154–155, 184, 189–190, 195, 197; edge effects, 43; fragmentation, 39, *39*; protection needs, 24–25; roadkills, 44, 191; travel problems of, *102–103*
Lion, mountain (*Puma concolor coryi*), 99
Lisbon, Portugal's riparian corridor plan, 259–261, *260*
Livestock: edge effects and, 43. *See also* Grazing
Loblolly pine seed predation, 159
Logging: conservation concessions, 249, 250; fire-suppression and, 34, *34*; overview, 13–14; regulations, 14; selective logging effects, 35. *See also* Clear-cutting; Deforestation
Lynx (*Lynx Canadensis*) and roads, 191

MacArthur, R.A., 50
Maine Wildlands Network, 106
Management: as adaptive, 240, 270–271; conservation agreements and, 244, 247–248; costs, 243–244; for small populations, 73
Management agreements, 247–248
Marsupial: *Antechinus agilis*, 109; *Hemibelideus lemuroids*, 71, 166
Marten: *Martes Americana* matrix resources, 133; *Martes* sp. corridor use, 185
Masai people, 257–258, 259
Masoala project in Madagascar, 217
Massey Creek project, Australia, 252
Matole River watershed group (California), 29
Matrix: communities and, 116–117; complexity of, 117, 121; as dynamic, 117; effects on population dynamics, 131, 133–143, *136, 138, 139, 141, 142*; levels of permeability, 121–123; Levins's model of metapopulation, 58, 117; metapopulation theory and, 58, 62, *63*, 131, 133; modeling challenges, 117; overview, 116, 145; quality, 117; size, 117
Matrix-edge effect: definition, 129; description, 129–130, *130. See also* Edge effects
Matrix effects on population dynamics: dispersal sink, 137–139, *138, 139*; exotics and, 143–145; matrix as resource,

133–134; overview, 131, 133; ROMPA, 136–137, *136*, 138; secondary habitat, 134–137, *136*; source habitat, *139*; spillover exploitation, 140–143, *141, 142*; "stopper," *138*, 139–140
Matrix traveling: asexual vs. sexual species, 124; barriers, 121, 122; distance-permeability, 122–123, *122*; home range, 122, *123*; knowledge on, 125; local extinctions, 123; overview, 121–125, *122, 123*; reasons for, 121; species variability in, 123–125
Measles critical community size, 84, 159
Mesoamerican Biological Corridor, 104
Metacommunity: connectivity level for, 82; corridors role, 84; debate on conservation approach, 84–85; description, 80–81; extinction and, 82, *83*; fragmentation of, 81; parts of, 81–82; patch isolation, 84; patch quality, *83*; patch size, 84; processes, 82, *83*, 84–85; seasonal fluctuations, *83*, 84; trophic levels and, 82, *83*, 84; types, 81
Metacommunity theory, 80–82, *83*, 84–85
Metapopulation: corridors and unusual processes, 161–162; definition, 56; distribution of, 56–57
Metapopulation Ecology (Hanski), 64
Metapopulation theory: categories of metapopulation, 59–60, *60*; conceptual history, 56–62, *60*; extinction and, 57, 58, 59–60, 62; Levins's model, 57–59; matrix and, 58, 62, *63*, 131, 133; parts summary, *63*; processes summary, 62, *63*, 64. *See also specific processes*
Microclimate alteration, 41–42, 154
Migration: corridors for, 101, 104, *105*, 108, 155; corridors for salmon, 26; definition, 64–65; Kwa Kuchinja Corridor, Tanzania, 257–259, *258*; lemmings needs, 134; *Oryzomys palustris* needs, 134; pronghorn antelope corridors, 101, 104, *105*, 155; sunfish needs, 134
Millennium Ecosystem Assessment, xii
Minimum threshold densities: antiregulating influences, 72–73, *74*; social behavior and, 77
Mining, 14
Mink (*Mustela vison*)/fragmentation, 201
Missing Linkage project, 179
Mission Mountain-Bob Marshall wilderness connectivity, 265